HISTORICAL CATASTROPHES: SNOWSTORMS & AVALANCHES

AN ADDISONIAN PRESS BOOK

by Walter R. Brown and Norman D. Anderson

HISTORICAL CATASTROPHES:

SNOWSTORMS & AVALANCHES

Titles in the Historical Catastrophe Series

Text Copyright © 1976 by Walter R. Brown and Norman D. Anderson
Illustrations Copyright © 1976 by Addison-Wesley Publishing Company, Inc.
All Rights Reserved
Addison-Wesley Publishing Company, Inc.
Reading, Massachusetts 01867
Printed in the United States of America
WZ/WZ   9/76   00828

Library of Congress Cataloging in Publication Data
Brown, Walter R   1929-
    Historical catastrophes—snowstorms and avalanches.

    "An Addisonian Press book."
    Includes index.
    SUMMARY: Discusses avalanches and snowstorms, citing
examples that constituted natural disasters. Final
chapter is devoted to the science of winter storms.
    1. Natural disasters—Juvenile literature. [1. Avalanches.
2. Blizzards. 3. Snow] I. Anderson, Norman D.,
joint author. II. Title.
GB5019.B76   973   76-11013   ISBN 0-201-00828-9

CONTENTS

CHAPTER ONE

TRAGEDY AT THE KNICKERBOCKER THEATER

THE MAN HAD JUST FINISHED feeding the lions when he saw the two people coming toward him. The two dim figures wading through the snowdrifts were about all that he could see—it was snowing so hard he had trouble seeing anything else. Both of them were dressed in khaki clothing, tall leather boots, and had broad brimmed hats pulled down over their eyes.

"Hello, there!" one of the people called to him. "Can you tell us how to get out of the park?"

"Can you believe it?" the other person said, laughing. "We are lost in a snowstorm in the middle

7

of the Washington zoo!" It was a woman's voice, and the words were spoken with a heavy European accent.

The zoo keeper gave the couple directions and watched as they trudged off. Dick Richardson held his wife's hand to keep her from slipping in the deep snow.

"It's getting dark," he said. "Time to go home, Bob."

"I wish you wouldn't call me *Bob*. That's a silly name for a girl."

"I know," he said, laughing. "But who can pronounce your real name?"

"Boukje is a perfectly good name," she insisted. "I know a lot of women named Boukje."

"In the Netherlands, perhaps," her husband replied. "But not in Washington, D.C."

The couple had been married for only seven months. In spite of the depression that the country was suffering through, Dick had managed to find a job as an electrical engineer. His new wife, Bob, was going to school and also worked part-time at the Embassy of the Netherlands as an interpreter and a secretary. Even with two salaries, they had little money. Because of this, they were living with Dick's parents in a large apartment a few blocks south of the zoo.

"I wonder if the Nelsons have arrived. Perhaps the snow kept them home."

"I doubt it, Bob. They could easily walk. And the way my parents like to play cards, I expect they insisted that they come, even in this snow."

It had been snowing in Washington for nearly 24 hours. Everything in the city was covered with a thick, wet blanket. More than two feet of heavy snow lay on the level areas. The limbs of the many trees that lined 18th Street were covered with white. The falling snow made halos around the dim bulbs in the street lights at each corner. Traffic had been stopped completely for hours, and the only sounds were the squeak of the Richardsons' shoes on the wet snow.

Even the streetcars had been stopped by the snow. Bob and Dick crossed and recrossed 18th Street, trying to find the buried tracks. But the snow was too deep.

The couple finally reached the apartment house and stamped their feet to shake the snow from their boots. They climbed the stairs and entered the door of the apartment. Four people looked up from their card game as they entered.

Bob and Dick greeted the older Richardsons and their guests. Then Bob gave a quiet signal to her husband. He followed her to the kitchen.

"Dick, I don't want to stay here while they play cards. We would be in the way. Let's go to the movies."

Dick quickly agreed. Living with his parents in the apartment was difficult for the newlyweds. They tried to avoid disturbing their parent's lives as much as they could. So they put on their coats, said goodby, and went back out into the snow.

Only four blocks away, at the corner of 18th Street and Columbia Road, stood the Knickerbocker Theater. The building was a modern one, made of brick, marble, cement and steel. The theater was only six years old. This was late January, 1922, so none of the moving picture theaters in the United States were very old. The oldest, in New York City, had been open for less than 20 years.

The picture that was being shown was called *School Days*. Along with it was a short comedy, *Get-Rich-Quick Wallingford*. Both were in black and white. There was some talk of someone trying to make a feature-length film in color, but no one in the small crowd that entered the Knickerbocker Theater that Saturday night had ever seen one.

As Dick bought the tickets, Bob looked at the deserted streets. The theater was shaped almost like a triangle, fitted into the odd angle made by the intersection of Columbia Road and 18th Street. The brick wall that ran along the 18th Street side of the building was straight, but the wall facing Columbia Road was curved. This curving wall matched the curve of the road and the sidewalk.

Bob remembered that another streetcar line ran down the center of the curving Columbia Road. Huge inter-urban cars, nearly three times as heavy as regular streetcars, often ran down these tracks. She and Dick had often ridden the rattling, shaking electric cars out into the country for picnics or to visit relatives who lived in nearby towns. Now the tracks lay deeply buried. The snow had even stopped the traffic on Columbia Road.

As a matter of fact, the snow had almost paralyzed the entire city. Even the fire department equipment was having trouble getting through. The hose wagon and the pumper truck that made up Engine Company 21 were both stuck in a snow bank only two blocks from their station house on Adams Mill Road. A young fireman, named Bill Hoeke, sweated under his heavy clothing as he dug and pushed to free the trucks.

"Rough going!" one of the firemen said, straightening the kinks in his back.

"Man, that bed is going to feel good tonight!" another answered.

The four firemen and their chief, Captain T.B. Stanton, had been called out several hours before to fight a fire in a chimney only a few blocks away from their station. The fire had not been a bad one but it had been tricky, and the men were tired and dirty when they started back to the station. Because of the snow, it had taken them almost an hour to reach the driveway of the firehouse.

"This is the roughest day I *ever* had," one of the older men complained.

But, as the trucks finally rolled into the station at eight o'clock, the long, terrible night was just beginning. It was to be the worst experience any of the men in the Company would ever have.

As Bill Hoeke carefully hung up his fire-fighting coat and hat and placed his boots under them, he wondered if he had made the right decision when he joined the Washington, D.C. Fire Department He had graduated from Catholic University just a few months before with a degree in mechanical engineering. But the big depression was at its worst, and he had not been able to find a job. Since he needed work, he joined the fire department. Everyone had told him that it was a mistake. College graduates did not work as firemen in 1922. In fact, very few high school graduates took jobs as firemen back then. The job required very little knowledge of anything, much less an understanding of engineering.

But it was a steady job, and Hoeke was young and strong. During the summer months, the work had been exciting and almost fun. The winter weather had made it more difficult and much more dangerous. Hoeke often wondered, especially on Saturday when he worked for 24 hours straight, when he would be able to quit and get a decent job.

Engine Company 21 had been busy during the past week, as it always was during cold weather. Fires in chimneys and flues were common, and several houses had caught fire from open fireplaces that were still common in Washington. For several days now it had been very cold. But the sky had been clear, and the people who lived in the Nation's Capital had enjoyed skating on the frozen water of the reflecting pool in front of the Lincoln Memorial.

The cold air poured down into Washington from the north. It was pushed there by a huge high pressure area that lay over Canada. Frigid air over the northeastern United States dropped temperatures into the low 20's over Washington, D.C.

To the south of Washington, off the coast of Georgia, was a low pressure area. The air in this low was warmer and contained a lot of moisture gathered up from the Atlantic Ocean. The mass of wet air slowly drifted northward.

The cold, dry air from Canada and the warm, wet air from the south met almost directly over Washington, D.C. The result was one of the heaviest snowfalls to have ever struck the Central Atlantic Coast. By Saturday afternoon, January 28, 1922, more than 25 inches of snow had fallen. This was more than twice as much snow as had ever fallen there before in a single 24-hour period.

As the Richardsons entered the theater, the comedy "short" was playing. Like all movies in 1922, it was silent. An orchestra provided music that helped set the mood of the picture. It was dark in the balcony where Bob and Dick liked to sit. They moved carefully down the aisle toward the box seats that lined the edge of the balcony.

"The man at the ticket booth said we could sit in the boxes if we wanted," Dick told his wife. "The theater has only about 200 people in it."

"Everyone must be sitting downstairs. I can see only four other people up here."

"Let's sit just in back of the box seats," Dick suggested. "I don't feel right about sitting in an expensive seat if I didn't pay for it."

Two small boys sat alone in the first row behind the box seats. The Richardsons felt their way into the row just behind them. The boys were laughing loudly at Bill Haines' antics on the screen. The couple sat down, and slowly their eyes became accustomed to the dim light.

"Move over one seat, so I can put our coats on the aisle," Dick suggested to his wife. She did not answer. He looked at her and started to speak again. But he stopped as he saw the look of horror on her face. He followed her gaze to the roof of the theater. As he did, she screamed.

A huge crack was racing across the ceiling. Plaster showered down on them. Then powdery snow blew in through the hole in the roof. With a deep rumble, the entire ceiling collapsed on the theater below.

"Get down!" Dick ordered, pushing his wife onto the floor between the seats. As he fell on top of her, the entire building shook and shivered as huge blocks of cement smashed downward.

Tons of debris fell onto the balcony. The backs of the seats kept it from crushing the Richardsons. But the weight caused the box seats and the first row to tear loose. The two boys who had been sitting just in front of the Richardsons screamed as the edge of the balcony fell onto the more than 200 people sitting on the ground floor.

The theater organist had been in the lobby of the theater trying to make a telephone call. He kept glancing at his watch as he listened to the strange tone made by the telephone. His call was not going through. In another minute or two, the comedy would be over, and it would be time for him to go to work playing music while people entered and left the theater. In disgust, he hung up the telephone receiver and opened the doors to the theater auditorium just as Bob Richardson screamed.

The organist stood in the open doorway in shocked horror. As he watched, the theater collapsed in front of him. He saw a huge chunk of cement from the ceiling smash down into the orchestra pit, killing the leader and several of the musicians. He turned and staggered into the snow-filled street.

Across the street stood a small shop with a light shining from its window. The organist ran through the snow toward the light. He pounded on the door and shouted. The woman who answered the door was an Italian emigrant who spoke very little English. She could not understand exactly what had happened. But she knew something horrible had frightened the man who stood shouting at her. Realizing that he could not make the woman understand him, the dazed man rushed away into the storm. The frightened woman ran into the street. She saw the fire alarm box on the pole at the corner. Not knowing what else to do, she pulled the alarm.

Bill Hoeke and the rest of Engine Company 21 had gone to bed. They had been back at the station for only an hour, but they were so tired that every man was asleep, except for the man on watch.

"Ping!"

The single stroke of the "joker bell" announced that a call was coming in from a fire box. Every man jumped from his bed. Slipping into their unlaced shoes and pulling up their suspenders, they ran for their trucks. Meanwhile, the tape punched its way out of the signal machine. The man on watch ran the tape quickly through his fingers, counting the holes that had been punched in it.

"Box 817," he called. "Box 817! Corner of 18th and Columbia Road!"

But every man in the Company knew where Box 817 was. It was only two blocks away in an area that was mostly homes and apartment houses. But near the box were also many old buildings with oiled, wooden floors that would burn like paper. And, of course, there was the big Knickerbocker Theater.

Captain Stanton hurried his men onto their trucks. The men of Aerial Ladder Company 9 were all ready to move, the engine of their ladder truck sputtering in the cold air. The doors were being opened. Everyone hurried. Box 817 would call out four engine companies, two ladder companies and the 4th Battalion Chief. They all knew that this could be a big fire.

The hose truck of Engine Company 21 was the first to hit the snow in the street. It carried 1,200 feet of 2½-inch hose, two tanks of soda-acid chemicals that could be turned on small fires through a ¾-inch hose, and all the axes, crowbars and other tools needed by the Company. The pumper truck followed close behind the hose truck, for without the added water pressure from the pump, the hoses would be of little use against a really big fire. Third in line was the ladder-carrying truck of Company 9.

The hose truck skidded around the corner onto 18th Street. Every man searched the buildings as they passed, looking for the glow of flames. Everything seemed quiet and peaceful as they approached the Columbia Road intersection. Suddenly the front wheels of the truck struck something in the snow. The driver slammed on the brakes. Bill Hoeke jumped from the truck to see what they had hit.

"What is it, Bill?" the driver yelled through the swirling snow.

"I don't know," Hoeke shouted back. "Big metal things. They look like brass doors!" He looked around him at the buildings on the corners of the intersection. "They are the doors of the theater! There must have been an explosion in there that blew the doors clear off their hinges!"

Without waiting for orders, the men of Engine Company 21 rushed into the lobby of the building. No one was in sight. The lobby seemed normal. There was no sign of smoke or of damage from an explosion. For some strange reason, it was very, very quiet.

"Check the furnace room," Captain Stanton ordered. "It must have been the furnace."

Bill Hoeke was dragging the nozzle of the ¾-inch soda-acid line, so it was his job to search for the fire. With one other man, he hurried down a short hallway to the furnace room door. The door was slightly open with one hinge broken off. A thin wisp of smoke crept through the crack. By the light of a flashlight, the two firemen stumbled down the stairs.

The fire was a small one. The sides of the wooden coal bin were smoldering, and hot coals lay all over the floor. A single burst of soda-acid foam put the fire out, and the two men returned to the lobby. (Later Hoeke was to figure out what must have happened. The rush of air caused by the falling ceiling must have blown down the vents and into the furnace, blowing the coals out into the room.)

"No explosion down there, Captain," Hoeke reported. "Only a small fire. We took care of it. What do we have up here?"

"We don't know. Something is blocking the doors into the auditorium. We can't get them open. Here comes the Ladder Company. *Get the Johnson Door Openers!*" he shouted.

The firemen from the Ladder Truck Company quickly popped one of the doors loose and pulled it away from its frame. A pile of rubble poured through the opening and onto the lobby floor.

"Oh, my God!" someone exclaimed. "The roof has fallen in!"

The pile of debris was topped by solid blocks of cement 18 inches thick. Under this were loose pieces of cement and chunks of plaster. Steel girders could also be seen here and there. The other doors were quickly ripped from their hinges. Firemen began to dig with their bare hands into the loose rubble.

"Here's a hand!" someone shouted. "Help me dig!"

"I've got someone here!" another voice called. "I need help! Get something to dig with!"

At that moment, the Chief of the 4th Battalion arrived. He quickly surveyed the situation. He grabbed at a passing fireman.

"Pull a 2nd, 3rd, 4th and 5th alarm," he ordered. Turning to Captain Stanton, he explained, "We'll need all the men we can get. Most of the engines won't be able to get through the snow. Get to a phone and call the police. Tell them to try to find more men and equipment. Suggest that they notify the Army and the Navy. We'll need all the medical help we can get, too, just in case anyone is still alive in there."

Bill Hoeke and another man had managed to dig into the loose rubble. As the tunnel grew, they took turns sliding in under the tons of cement that topped the pile of debris. One man would slide into the hole, dig away at the plaster and cement with a crow-bar, then call to be pulled out. Soon Hoeke found a body in the tunnel. It was a small boy. Hoeke felt the limp wrist and found a pulse.

"I've found someone," he called back. "And he is still alive!"

Carefully he dug and dug until he could pull the boy free. Then he called back into the lobby.

"I've got him. Pull me out. But do it slowly."

The other fireman grabbed Hoeke's heels and pulled him gently backward. Carefully the man and the boy were dragged out into the lobby. Within a few minutes, the sobbing boy was telling his story. He had been sitting behind the box seats, in the first row of the balcony, he said, when he heard a woman sitting behind him scream. Then the roof had fallen in, and the balcony had given way. His friend, he told them, was still buried in there somewhere.

From where the firemen stood they could tell that not all of the balcony had collapsed. The Chief turned to the fireman nearest him.

"You," he said, pointing to Bill Hoeke and his friend. "Tell the Ladder Company men to pop the doors off in the back of the building. See if you can get into the auditorium that way. If you can, check the balcony and see if anyone is still up there."

The firemen quickly did as they were told. They found the doors in back of the stage unblocked and easily made their way into the theater. In the auditorium, they found a pile of rubble higher than a man's head. They shone their lights up toward the ragged edge of what had been the balcony to see if anyone was there.

"Hey, down there!" a man's voice called from the darkness. "Up here! On the balcony!"

In the dim light the firemen could see four people. They were sitting calmly on top of the collapsed ceiling with the snow falling all around them through the hole in the roof.

"Are you all right?" Hoeke called.

"We're okay," Dick Richardson answered. "My wife and I have a few cuts and bruises. The backs of the seats kept most of the ceiling off of us. Can you get us down?"

The firemen quickly found a long pole, which they leaned against the shattered edge of the balcony. Bob Richardson quickly wrapped her legs around the pole and slid down. Her husband and the other couple soon joined her on the smashed cement.

"Engine Company 21! Ladder Company 9!" someone was calling as the little group of survivors made its way from the building. "Hurry it up, men! We've got another call."

Hoeke and the other men of Engine Company 21 hurried to their trucks. It was 11:25 and the snow was still falling in the dark night.

"We have a call from Box 245," the men were told. "An apartment house fire."

"Hey, here come the Marines!"

As the three fire trucks rounded the corner, a detachment of Marines marched double time up the street. The men were sweating in the cold air after a two-hour forced march through the knee-deep snow. A mile or so behind them, crews from docked Navy ships were riding in electric trucks that could get through the snow almost as easily as a man could. Behind them, at Fort Meyers, ambulances were being hitched to teams of horses.

It was dawn before the men of Engine Company 21 got back to the Knickerbocker Theater. They had fought three different fires in the past eight hours. Every man was dead tired, but no one suggested that they should go back to the firehouse. Throughout the long night they had thought about the people trapped in the theater.

"I sure hope those Navy boys know what to do," someone said. "There's nothing in our manual about how to lift 10 tons of cement off someone!"

The military personnel sent to the disaster scene did know what to do. Word was quickly sent back to the Navy Gun Works. Soon teams of mules pulling heavy wagons were on the streets of Washington. On the carts were huge hydraulic jacks, which were used in the Gun Works to lift and support large gun barrels.

Once these jacks were in place, the edges of the cement blocks were slowly and carefully lifted upward. With a few inches of extra space in which to work, men could now safely slide under the cement and dig for the trapped people. Some, like the Richardsons, were found with only minor injuries. But more than half of the people in the theater were either dead or dying.

The rescue work continued all through the next day and far into the following night. The injured were taken to nearby homes. The dead were taken to a church a half a block away. The final death toll was close to 100 people.

No one ever knew exactly why the roof of the new theater collapsed. The snow was very wet and more than two feet of it had fallen on the roof. But even this tremendous weight should not have been enough to cause the disaster. It was suggested that the curved wall of the building, on the Columbia Road side, had given way enough to let the roof fall. It was also suggested that the steady vibration of the huge inter-urban cars crossing at the intersection had weakened the building.

What happened to those people who met briefly in the ruins of the theater? The Richardsons moved to Florida, just in time to be caught in the terrible hurricane of 1926. Bob Richardson remembers sitting on her kitchen table, with sea water sloshing all around her, and thinking that another roof was going to fall on her. But this time, her story was different. The roof blew away, along with many of her belongings.

Bill Hoeke stayed with the fire department and eventually became a Fire Commissioner in Montgomery County, Maryland, a suburb of Washington. He answered many calls for help during his long career, but he still feels that the terrible night at the Knickerbocker Theater was the worst.

For 53 years Bill Hoeke didn't know the name of the woman he had rescued from the balcony of the theater. And Boukje "Bob" Richardson didn't know who her rescuer was. Then, in 1975, the two met again, quite by accident, less than 200 miles away from the site of their common adventure. You can imagine the time they had comparing notes on the tragedy at the Knickerbocker Theater.

CHAPTER TWO

THE GRANDDADDY OF ALL BLIZZARDS

"MY, THAT WAS a fine meal," Grandfather Taylor sighed as he took one last bite of roast beef.

Six people were seated around the table in front of the large window overlooking Central Park. They were just finishing their Sunday dinner and were about ready to have their dessert.

Grandfather and Grandmother Taylor had arrived from Florida on Friday at the height of a snowstorm. They were here to visit Lisa and Bill Mills and their grandchildren Dorothy and Donald. There also was a very important meeting in New York tomorrow, Monday, March 12. Neither of them intended to miss it!

"This is some weather," Mrs. Mills apologized to her parents.

"It's great," waved Grandfather Taylor. "Besides being away from you folks, the only thing we miss in Florida is the snow. Besides, a good snowstorm is the best way I know to celebrate the 40th anniversary of the Blizzard of '88."

Dorothy and Donald had not seen their grandparents for several years, and they had been looking forward to this visit. They especially wanted to hear their grandparents' firsthand account of the great blizzard of 1888.

"Why don't we move into the living room where we will be more comfortable?" Mrs. Mills asked.

"Fine," her father responded. "Why don't you get the scrapbook?" he added to his wife. Over the years the Taylors had collected newspaper clippings, magazine stories, photographs and other items about the Blizzard of '88.

When they were all seated, Mr. Taylor began. "We were living in an apartment not too far from here. Your Grandmother and I had only been married for about a year. I was working for an investment bank on Wall Street at the time.

"There were no automobiles, of course, in 1888. I used to go back and forth to work on a streetcar. We took the train when we wanted to go someplace out of town.

"I remember waking up sometime near midnight on Sunday. Outside it was raining 'cats and dogs' as we used to say. The rain suddenly changed to sleet and then to a needlelike snow.

"By morning enough snow had fallen to make the streets almost impassable. Shortly after seven o'clock the Brooklyn bridge was closed to pedestrians. One man insisted he had to get from Brooklyn to Manhattan. He ended up having to be rescued by the police before he was half way across. The fierce winds blew the snow with such force that people were swept off their feet. Often you couldn't see more than a few hundred feet ahead.

"I hadn't missed a day of work in my life, and I was determined that March 12, 1888, was not going to be an exception. The streetcar I usually rode on was a cable car. However, the cars weren't running that Monday morning. The ice and snow on the tracks made it impossible for the grip to reach the cable. The snow already was two feet deep, and in some places the winds had piled up drifts higher than a man's head.

"Luckily I managed to catch one of the few horse-drawn streetcars that were operating. However, about half way to work the car got stuck in a large drift. The driver and conductor unhitched the four horses and set off on horseback for the streetcar stable. We passengers were left behind and had to figure out for ourselves what to do next.

"The only thing to do was get out and walk. There were only a few hardy souls on the street compared to the usual Monday rush. I ducked into a building every few blocks to warm myself and catch my breath. Other people had the same idea, and soon there was a group of a half dozen or so headed toward offices downtown. We were all strangers, but in the face of the storm, we soon became friends.

"There was a spirit of adventure in the air. People were quick to give each other a helping hand. Policemen braved the cold and blowing snow to aid those who were on the streets. In front of one large hotel, several officers were busy rubbing snow on the frost bitten ears of some of the pedestrians."

"Did you make it to work, Grandpa?" Donald interrupted.

"Yes, but not many others did. When the gong sounded at ten o'clock to begin trading at the Stock Exchange, there were only 30 of the 1,100 members on the floor. As time wore on, the crowd did not increase much. The floor looked like a deserted ballroom.

"Only three out of the 30 to 40 banks, which did business for members of the Stock Exchange, were represented. Little business could be conducted, and so at noon the Exchange closed for the day. Nothing like that had ever happened before in Wall Street, although a big sleet and ice storm in 1881 did delay business when the telegraph lines broke and the tickers quit working."

Mr. Taylor paused to accept a cup of coffee from his daughter. Dorothy wanted to know how he got home that snowy Monday night.

"We didn't," Mr. Taylor answered. "Several of us slept on our desks in the office. In fact, people slept almost everywhere. Macy's Department store spread mattresses on the floor for their employees. Many sat up all night in hotel lobbies and hallways.

At least one railroad rented Pullman Berths in snowbound railroad cars at $2.00 apiece.

"The city was completely isolated by Monday afternoon. No trains were entering or leaving the city. Telephone and telegraph lines were down. New York City was completely cut off from the rest of the country. It wasn't until Wednesday that contact was made with Boston by way of London. The underseas cables continued to operate and bring us news of Europe, even though we didn't know what was happening a hundred miles away.

"In spite of our being stranded, there was surprisingly little panic and not too much hardship. A few fist fights broke out when a tavern keeper or grocery store owner raised his prices to take advantage of the situation.

"Some people, of course, profited from the blizzard by shoveling snow and running errands. The son of a man in our office made a small fortune in a rather unusual way and because of a rather strange set of circumstances.

"It all began when a huge ice floe came down the Hudson River. On Tuesday morning it was in the bay off Governor's Island. When the tide turned, it floated up past the Battery into the East River. It finally became sandwiched between the shores of Manhattan and Long Island.

"The ice was at least six inches thick. It was covered with an additional two inches of hard snow. The floe, which was solid from shore to shore, blockaded the Fulton and Wall Street ferry slips on both sides of the river.

"All of this happened at a time when the entrances to the bridges were jammed with hundreds of people. They were surging and crowding for a chance to get across the river by bridge cars. When several in the crowd saw that it might be hours before they could cross the bridge, they set out on foot for the ferry landings. They didn't realize that the ferries had been stopped by the ice floe."

"Couldn't they walk across the river on the ice?" Donald asked.

"That's exactly what happened. Some of the people made their way along the shore to a dock near Martin's store on the Brooklyn side. They stood looking down at the ice, not sure if they should try it.

"It was at that point that my friend's son came along with a ladder. The boy had a shrewd eye for business. He planted the ladder on the ice and scrambled down from the dock. Then he jumped up and down on the ice several times. This proved to those watching from the dock that it was strong enough to walk on.

"Many people scrambled down upon the ice, each paying the boy two cents for the use of the ladder. They moved out cautiously, still not completely sure that the ice would support their weight. When they reached the end of the dock, a swirling blast of wind knocked several of them off their feet. They helped each other up and set out again, determined to reach their destinations.

"They finally made their way to a landing at the foot of Beekman and Fulton Streets. They called to a young laborer at a fish market nearby for help. They had no way of getting up off the ice. The young man found a ladder, but he was not to be out-done by the young boy on the Brooklyn side. He charged each person five cents for the use of his ladder.

"This went on for an hour or so with several hundred people making the crossing. Then the Brooklyn police, fearful of a catastrophe, stopped any more people from going onto the ice.

"At about the same time, two large tug boats came steaming up the river at full speed. They rammed their sharp steel prows into the ice floe. They hoped to free the ice jam so the ferries could start running again. Jagged lines appeared in the ice the instant contact was made.

"There were still dozens of people on the ice when the tugs began their assault. Seeing what was about to happen, they scattered and ran for the nearest shoreline like a flock of frightened sheep. Nearly all of them had reached shore when the floe, with a noise like a small cannon, parted and began to move."

"Do you mean there were still people on the ice when it started to break up?" Dorothy wanted to know.

"Yes, there were at least a half dozen who had been in the middle of the river when the tugs rammed the floe. All means of escape were cut off. However, the crew on the larger of the two tugs quickly saw the stranded people's predicament.

"The large tug reversed its course and slowly

moved up to the edge of the floe. It took several tries before the crew succeeded in throwing a rope to the stranded men. Eventually all of the men were taken aboard the tug, only a little worse for their adventure."

"I don't think I would want to try crossing the river on the ice if it ever freezes over," Donald volunteered.

"Little chance, Donald. We just don't have winters like we used to," replied Grandfather Taylor, continuing his theme that the Blizzard of '88 would never be matched.

"Grandma, where were you when all of this happened?" Dorothy wanted to know. "Were you home all by yourself?"

"No, I had gone to Albany with a friend several days earlier to see my mother," Mrs. Taylor replied. "As things turned out, we had every bit as much of an adventure as your grandfather."

"We were on our way back to New York on the Albany Express when the storm struck. It was dark outside, and we had trouble seeing how hard it was snowing. However, when we passed through a town, the lights showed us that the snow was getting deeper and deeper. Finally the train came to a stop somewhere above Hastings.

"The engines were disconnected so they could be taken into Hastings to get water for their boilers. However, they stalled in a big snowdrift a short way down the track. They had to wait there until they were dug out by snowplows working their way northward.

"We were fortunate that the cars were not heated by steam from the engines. Instead, each of the cars had its own stove or heater, and we were kept comfortably warm. The railroad company sent us cans of milk, sandwiches and pies from a nearby village. With plenty of food and fuel, we were in no real danger.

"Many of the passengers played cards and dominoes to help pass the time. Others napped, and a few devised practical jokes to relieve the monotony aboard the stranded train. One of the passengers, from somewhere out in the midwest, kept saying that this storm was nothing compared to the blizzards on the western plains. He kept it up until a fight broke out.

"Your grandfather and I learned many years later that an Iowa newspaper editor in 1870 is believed to be the first one to use the word *blizzard* to describe a severe snowstorm. Originally the word was used to mean a volley of shots. If you had ever been trapped in a blizzard, you would know that it really seems as if Mother Nature is firing snow at you."

"What else happened on the train?" asked Donald, who wasn't too interested in Grandmother Taylor's language lesson.

"Well, for one thing, about twenty of the gentlemen passengers formed an Association which they called *The Snow Birds.* They decided to have a dinner each year on March 12 at the International Hotel in Hastings to celebrate our being snowbound."

"You mean they were like the *Blizzard Men of '88* here in New York?" Mr. Mills asked.

"Yes, but I don't know if they went through with their meetings," Mrs. Taylor replied. "One other thing though—most of the passengers on the train joined *The Snow Birds* in signing a resolution. I have a copy of it in our scrapbook. Here it is."

Whereas, This organization, as well as a large number of fellow-passengers, became stalled on the New York Central and Hudson River Railroad between Dobbs Ferry and Hastings-on-Hudson, N.Y., on March 12, 1888, and

Whereas, The said company did voluntarily assume the responsibility of paying for our maintenance and also securing everything in its power to make its patrons comfortable, therefore be it

Resolved, That the thanks of this organization are due and are hereby tendered to the said New York Central and Hudson River Railroad and its employees for the prompt and kind attention in providing for our physical comfort and in procuring hospitality for the night;

Resolved, That a copy of these preambles and resolutions be sent to the New York Central and Hudson Railroad, and that the same be entered in full upon the minutes of this organization.

"When did you finally get home?" asked Donald, who wasn't interested in the faded newspaper clipping and the strange language of the resolution.

"By Tuesday afternoon the track from the snowbound train to Hastings was cleared. They took our train there, and many of us spent the night in private homes. On Wednesday we finally made it into the city. Our train was the first one to reach Grand

Central Station on the tracks of the New York Central and Hudson River Railroad.

"My friend and I hired a horsedrawn sleigh to take us to our apartment. I never told your grandfather how much it cost for that short ride. However, I guess it would not have mattered much. We were both overjoyed to be together again."

"Can we look at the other things in your scrapbook?" Dorothy asked.

The first thing that caught Dorothy's eye was a newspaper clipping describing the cause of the storm. Information about the weather in 1888 came from a small network of stations run by the United States Army Signal Corps. There was no weather service as we know it today.

In New York City, the weather station was in a room on the top floor of the Equitable Building. When the telephone and telegraph lines went down as a result of their heavy loads of ice and snow, the man on duty in New York City lost touch with the other stations. However, he still was able to piece together a good picture of what happened. On Tuesday he told a reporter:

"Yesterday's storm was really a combination of two storms in one: one winter cyclone from the west and another from the south. On March 8th the western disturbance was gathering strength in the neighborhood of Salt Lake City. On March 9 it was in Colorado. On March 10 it was moving along the Missouri Valley. Sunday night it was centered over the Lakes region and moving southeast.

"On Sunday another storm center was reported

in Georgia. It was swinging toward the northeast. These two big storms met off Cape Hatteras with a bang. They gave a big twist and made a beeline up the coast for New York and points north. Neither of these storms was an amateur. Both of them were deep lows and going strong when they centered smack over Manhattan between two and three o'clock Monday afternoon. The storm from the west had the Arctic on its heels. The storm from Georgia was sopping with warm moist air from the Gulf of Mexico. You couldn't find a better combination for concocting a howling blizzard!"

The big blizzard did not shut down New York City completely. Even on Monday night, a few hours after the height of the storm, a hundred people showed up for the circus at Madison Square Garden. The famous P. T. Barnum was there himself, and *The Greatest Show on Earth* went on. So did the plays at three Broadway theaters, although most of the seats were empty.

The mercury registered 1°F below zero on Tuesday morning. As might be expected, stoves overheated, and houses and other buildings caught fire. Firemen struggled to get their smoking engines and ladder wagons through the streets of the snowbound city. Fortunately no major fires broke out while the city was trapped under the snow.

By Tuesday afternoon, Broadway began to come alive with snow shovelers and sleighs. It seemed like every cutter and bobsled in the city had been brought out of storage.

(above) *New York City starting to shovel out after the Blizzard of 1888.*

(above right) *Burning holes in the snow was one way of clearing the streets. Woodcut from Harper's Weekly, March 24, 1888.*

(right) *Firemen struggling to answer an alarm during blizzard. Woodcut from Harper's Weekly, March 24, 1888.*

Wednesday was bonfire day. Merchants dug holes in the drifts and set wood afire to melt the snow. Others attacked it with streams of hot water. Barrels of rock salt were dumped on the streetcar tracks. Railroads hired snow shovelers at the unheard-of-price of $5 a day.

By Thursday the sun was shining, and the snow began to melt. Signs began to appear in snowdrifts. One on East 23rd Street, near Madison Square, read:

THIS SNOW IS ABSOLUTELY FREE!
PLEASE TAKE A SAMPLE!

One shopkeeper thought for sure that he had come up with a way to get his sidewalk cleared. His sign read:

IMPORTANT!
EXPENSIVE DIAMOND RING LOST UNDER
THE SNOWDRIFT!
FINDER'S KEEPERS!
START DIGGING! YOU MAY BE THE LUCKY
WINNER!

Unfortunately, he had no takers and finally had to clear the sidewalk himself.

The Blizzard of '88 hit Connecticut, western Massachusetts and the Hudson River valley even harder than New York City. Snowfall ranged from 40 to 50 inches. Drifts piled up to 30 or even 40 feet deep. In New Haven, Connecticut, on Whitley Avenue, there was one drift that measured 53 feet!

In Middletown, New York, citizens crossed the business streets through tunnels burrowed under snow drifts. The snow there reached above the second floors of houses and stores. The main street in Pittsfield, Massachusetts, lay under 20 feet of snow—some houses were entirely covered.

The Taylors' scrapbook contained a letter from a man who had grown up in the midwest. He was one of many who got tired of the New Yorkers claiming the Blizzard of '88 as their very own.

"The day of the blizzard I was a pupil in the sixth grade of the Plainview, Nebraska, public schools. The storm struck suddenly during the noon recess and thrilled the children. They raced wildly down the hill from the schoolhouse and then delighted in fighting their way back up.

"The fury of the wind increased until mid-afternoon. About that time some of the men of the town called at the school and conferred with the teachers about dismissing the pupils. It was finally decided to keep them at school if their parents did not call for them.

"My chum's father called for her and agreed to be responsible for me. On the way home we became lost. We experienced a great deal of suffering until we found ourselves back at the schoolhouse.

"Children in some of the other schools were not as fortunate. One teacher tried to lead her pupils home, became lost, and they all perished. Another, more fortunate, led her pupils two miles to safety. They still tell stories out here in Nebraska about the Blizzard of '88."

Perhaps one of the strangest items in the scrapbook was an editorial. It had appeared on Sunday, March 11, 1888, in the *News and Observer* in

Raleigh, North Carolina. This, of course, was the morning before the big storm in the northeast. The writer, it seems, was concerned about the fierce nature of the winters on the western frontiers. When he wrote the editorial, it probably never occurred to him that the nice safe eastern seaboard could have a howling blizzard. Nor did he dream it would happen the very next day.

The Lesson of the Blizzard

There is, as it seems to us, one lesson to be drawn from all of this horrible experience, and that is the folly of settling in a wild country. Especially on the bleak plains, however cheap the land may be, without assured means of providing decent shelter, fuel, and provisions for more than the immediate future. The representatives of land agents and other considerations have induced many to take up western lands without means, on the chance that the success of the first crop would enable them to provide adequately for their families. This is gambling in human life.

Another editorial, this one in the Tuesday edition of the *New York Times,* argued that New Yorkers could learn a lesson from the blizzard.

The blizzard yesterday may accomplish what months, if not years, of argument and agitation might have failed to do. Now things are tolerably certain—that a system of really rapid transit which can not be made inoperative by storms must be straight way devised, and as speedily as possible constructed, and that all the electric wires, telegraph, telephone, fire alarms and illuminating, must be put underground at once.

A lot of people agreed with the editorial. Within a year or two, telegraph and telephone wires were placed underground in New York and several other major cities. Plans were prepared for subways in New York and Boston which still are in use today. And other precautions were taken so the northeast would never again be crippled as it was in 1888. Of course there is little chance of this happening because, according to the loyal members and followers of the *Blizzard of '88 Society,* blizzards just aren't as severe as they used to be.

CHAPTER THREE

TRAGEDY AT DONNER PASS

As THEY RODE ALONG, they looked in fear at the mountains ahead. Although it was only the last week in October, snow was already falling. Winter was beginning almost a month earlier than usual. All the trails were covered with snow. Their only guide was the mountain tops, which it seemed they would never reach.

Twelve-year-old Virginia Reed could hardly remember that warm, sunny April day when she and her family had left Springfield, Illinois. At first the trip had been exciting. Just think—going all the way to California by wagon train!

The Reed family had started out with three wagons, each pulled by three yoke, or pairs, of oxen. The big family wagon had been especially built for the long trip. It was a two-story wagon which became known as the *Pioneer Palace Car*. The door to the wagon was on the side, like an old-fashioned stagecoach. Just inside the door was a small room. On either side of the room were spring seats with comfortable high backs. In the center of the little room was a small, sheet-iron stove. A circle of tin in the wagon's canvas cover kept the hot stove pipe from setting the wagon on fire.

Besides Virginia, there was her father and mother, and Patty, age 8, James, Junior, age 5 and Thomas, age 3. Also in the Reed party was Grandmother Keyes, a hired girl and her brother, plus three young men hired to drive the oxen. The Reeds were traveling with the two Donner families from Springfield and planned to join others when they reached Independence, Missouri.

The trip to Independence was a big adventure for Virginia and the other children. Virginia had been allowed to bring her pony, Billy. Part of the time she rode up ahead of the wagon train, one of the first to see what might be over the next hill. At other times she would ride along behind and help herd the cattle that the Reeds and the Donners were taking along to California.

Everything went well until they were part of the way across Kansas. Grandma Keyes was in good health, and the farther west they went, the more her spirits improved. Then they came to the Big

Blue River. They found it swollen by the spring rains, and there was no way across except on rafts which had to be built. As soon as the wagons stopped moving, Grandma Keyes's health began to fail. On May 29, 1846, she died.

After crossing the Big Blue River by raft, the wagons pushed on westward. They numbered about forty now, counting those wagons that joined the Reeds and the Donners in Independence. On a good day they made from 15 to 20 miles, shortening or lengthening the distance in order to obtain a good campsite for the night.

After several weeks they reached a place called Hasting's Cut-off. This was the beginning of a new trail that passed around the south end of the Great Salt Lake in Utah. According to Lansford Hastings, the explorer who had opened up the route, it shortened the distance to California by 300 miles, and the only bad part was the 40-mile stretch through the desert by the shore of the lake.

There was great debate about which route to take. Finally on July 31, 87 people, including the Reeds and the Donners, left the main party. They set off in high spirits on the Hasting's Cut-off.

A few days travel showed them that the new route was not as good as it had been described. Following the route Hastings had laid out, they soon came to rugged mountains which the wagons could not cross. A new route over the mountains had to be found and progress was slow. As it turned out, it took them a month, instead of the week they had planned on, to reach the Great Salt Lake.

Then there was the long trip across the desert. Hastings said it was only 40 miles. It turned out to be more than 80 miles. In crossing the desert, the Reeds lost most of their cattle and oxen when the animals wandered off in search of water. This meant packing all of their belongings into their smallest wagon and leaving everything else behind.

Upon reaching the western edge of the desert, an inventory of food and other supplies was made. It was clear that the party did not have enough food to last them until they reached California. And to make matters worse, the first storm of the approaching winter struck the weary travelers. Nearby hilltops were white with snow, and suddenly California seemed far, far away.

It was decided that someone must go ahead to Sutter's Fort on the other side of the Sierra Nevada Mountains and bring back provisions. William McCutchen and C. T. Stanton volunteered. They left in mid-September, carrying letters to Captain Sutter asking for help.

The Reed family and the others pushed on with broken spirits. Both people and animals were exhausted from the five months of hard travel. And disaster was soon to strike again.

Virginia's father got into an argument with John Snyder, one of the drivers, about the best way to handle the oxen. Tempers flared and heated words were exchanged. A fight broke out and Snyder struck Reed several times with his whip. Reed was stunned and partially blinded by the blood streaming from the gashes in his head.

When Mrs. Reed saw what was happening, she
ran between the two men. But Snyder could not
stop—the whip had already started its downward
action. Reed's knife was out of its sheath before the
first blow struck his wife. He jumped at Snyder.
The knife found its target. Snyder fell, fatally
wounded.

Snyder was buried the next day and a council
meeting followed. The council refused to accept
Reed's pleas of self-defense and ordered that he be
sent into the wilderness. At first he refused to go.
However, his wife pleaded with him to go, since she
feared he might meet with violence if he stayed.
She also suggested that, if he went on, he could
return later and meet them with food.

Reed finally consented to go when the rest of the
party agreed to take care of his wife and four children. It was a sad day when James Reed said
good-bye and set out alone on horseback. Little did
the Reed family realize that this misfortune would
later save their lives.

The party traveled on, but there was little happi-
ness left in the group. However, on October 19, the
Reeds and the others had reason to rejoice. Stanton
and seven mules loaded with provisions returned
from California. For the Reeds, Stanton brought
something even better than food—news that James
Reed was alive. Stanton had met him not far from
Sutter's Fort. Stanton had given him a fresh horse
and some food. By now he should be at Sutter's
Fort organizing a relief party to bring aid to his
family and the others.

Stanton helped the Reeds pack what little they had left on one of the mules, and they started out once more. Mrs. Reed, with Tommy in her lap, rode on one mule. Patty and Jimmy rode behind the two Indian guides that returned with Stanton from California. Virginia stayed with Mr. Stanton, thankful to be riding instead of walking as many of the others had to do.

Now, as they traveled westward, winter started in earnest. It was snowing harder, and as the snow got deeper, it became impossible for the oxen to pull the heavy wagons. What provisions they had left were unloaded from the wagons and packed on the backs of the oxen. Another start was made, with men and women walking in snow up to their waists. Some carried children in their arms; others tried to drive their cattle through the snow.

A halt was called when the Indian guides could no longer find the trail. Stanton and the guides went ahead to see if they could find the road. They came back to report that the party could make it now, but they must waste no time in getting over the summit just ahead. Most in the party were so exhausted that they refused to take another step. Those that favored a forced march to the other side of the mountain gave in. Camp was set up within three miles of the summit.

That night it snowed again. Great feathery flakes came whirling down. The air was so full of snow that only objects a few feet away could be seen in the light from the campfire. In the morning the snow lay deep on the mountains.

The group the Reeds were traveling with turned back. They sought temporary shelter during the first days of November in a cabin that had been built by another party trapped by snow two years earlier. Trees were cut down, and two more double-cabins were built. The old cabin became known as the *Breen Cabin,* since it was occupied mainly by the Breen family. The new ones were named the *Murphy Cabin* and the *Reed-Graves Cabin.*

The three cabins were located near a lake, which has since become known as *Donner Lake.* The Donner families had been traveling a few miles apart from the rest of the party for several weeks. They were now camped in Alder Creek Valley below the lake. The snow had come on so suddenly there that they had no time to build cabins. Instead they had hastily put up brush sheds, covering them with pine boughs. In many ways they were worse off than those camped up by the lake.

Most of the cattle were killed for food, and the meat was placed in the snow to keep it from spoiling. Mrs. Reed had no cattle to kill, but she promised the others two cows when they reached California for each one they would give her family now. The hides from the animals were stretched across the cabin roofs to keep most of the snow from drifting in.

At first the time went by rapidly. As soon as the storm let up, there was work to be done. Firewood had to be cut; the cabins had to be made more weatherproof; food had to be brought in, and dozens of other tasks needed to be taken care of. And plans had to be made on how they could get over the mountain before it was too late.

It was decided the only solution was for some of the strongest to try to escape across the mountain pass. This would leave more food for those that stayed behind. On November 12, the first clear day, 15 adults decided to try to cross the mountain. They made it over what was to become known as *Donner Pass* and spent the night camped just beyond. However, afraid that they couldn't make it, they returned to the cabins by Donner Lake on the second day.

A second attempt was made on November 21. However, the snow was so deep the group was forced to return before the day was over.

More snowstorms followed. Some lasted for several days. Others were periods of stormy weather lasting only a few hours or a day at the most. Patrick Breen recorded in his diary that by December 13, the snow was eight feet deep on the level ground.

On December 16, just nine days before Christmas, a group of 17 started out on a third attempt to escape. Two gave up and returned to camp the first day. The others, ten men and five women, kept on going. They carried only enough food to last them six days. It took them two days to get across the top of the pass.

They kept on going because it seemed useless to turn back. Later they were called the *Forlorn Hope* party. According to the dictionary, this name describes a group selected, usually from volunteers, to perform an almost hopeless undertaking. In this case the name certainly was appropriate.

By the ninth day the group ran out of food. A storm kept them huddled beneath a few thin blankets for the next two days and two nights. Four of them died. A fifth member had been left behind on the trail before the storm struck. The others could survive only by turning to cannibalism. Half frozen and starved beyond belief, the ten survivors ate their dead friends in one last effort to escape death.

Then they struggled on—cold, exhausted and weak from starvation. As they came down off the mountains to where the snow lay only in patches, they were lucky enough to kill a deer. But the food was too late, and three more of the party died. The remaining seven, two men and five women, struggled on until they came to a village of friendly Indians. From there they were taken to a ranch in the Sacramento Valley where they were nursed back to health.

While the Forlorn Hope party was battling its way across the snowy mountains, Christmas came and passed at the cabins on Donner Lake. Virginia could hardly remember what Christmas had been like a year ago in Illinois. It seemed so long ago and so much had happened since. She had even more trouble imagining what Christmas would be like next year. Would they all be together in warm and sunny California? She didn't want to think about the other possibilities.

Mrs. Reed had decided weeks earlier that her children would have a treat on the special day. She had laid away a few dried apples, some beans and a small piece of bacon. When the children saw what she had for them, they jumped up and down with joy. The food was cooked carefully so not a scrap would be wasted. When they sat down to eat, Mrs. Reed said, "Children, eat slowly, for this is one day you can have all you want."

As 1846 ended and the new year began, the situation in the camps got worse. Most of the families had run out of beef. Many were trying to live on the cattle hides. If the hides were boiled long enough, they produced a gluelike soup. It tasted terrible but it gave a little nourishment.

When the food ran low at the Reeds, they killed their dog. Mrs. Reed and her four starving children lived on the dog for a week, eating everything but the bones. There were more deaths, and stories of people in the other cabins eating their dead relatives. Mrs. Reed was determined this would not happen to her family.

Meanwhile, at Sutter's Fort, work was under way to organize a relief party. Captain Sutter offered to do everything possible to help. He gave Reed and another man horses and provisions.

The two started out for the mountains and went as far as possible with the horses. Then they placed the provisions in backpacks and proceeded on foot. However, a storm set in, and they finally were forced to return to Sutter's Fort.

Captain Sutter advised them to go on to Yerba Buena, which is now San Francisco, for help. All of the able-bodied men at Sutter's Fort were in southern California fighting in the Mexican War. Reed took his advice and went to the naval officer in charge at Yerba Buena to plead for help.

Reed was at Yerba Buena when the seven members of the Forlorn Hope party arrived from across the mountains. News of their famished condition told the story of what it must be like at Donner Lake. Cattle were killed at once, and the men stayed up all night drying beef and making flour.

A party of seven, under the command of Captain Tucker, left as soon as enough beef and flour had been prepared. They had a long and difficult journey. They finally reached the cabins by the lake on the evening of February 18, 1847.

The cabins were almost completely covered with snow. The rescue party shouted to see if anyone was alive. Mr. Breen climbed up the icy steps from his cabin and shouted back, "Relief, thank God, relief!" His words echoed across the mountains and also were repeated a hundred times or more by the other thankful survivors.

There was food and rejoicing. There was news that other relief parties were being organized to bring them more food and to rescue them. And there was sadness, because several had died just days before the relief party had arrived. Among them was Milt Elliot, a faithful friend, who had seemed like a brother to the Reed children.

On February 22, a party of 23 started out in one more attempt to reach safety. In the party was Mrs. Reed and her four children. It was a bright sunny day. The good weather, plus the food brought by

Captain Tucker four days earlier, had put everybody in good spirits. However, they had not gone
far when Patty and Tommy Reed gave out, unable
to make their way through the deep snow.

At first Mrs. Reed said she would go back with
the children. The leaders of the party didn't think it
was a good idea and would not allow her to do so.
Finally, Mrs. Reed agreed to let Mr. Glover take the
two children back to the Breens' cabin. As they
parted, Patty said, "Mother, keep going and maybe
you will meet father. And don't worry, I'll take care
of Tommy."

The party went on. The men wearing snowshoes
broke a path, and the others followed in their
tracks. At night they lay down to sleep, only to awak-
en in the morning to find their clothes frozen. At
the break of day they were on the trail again.

They were able to make better time when the
snow was frozen. The sunshine, which at first
seemed to be such a blessing, only added to their
misery. The sun's heat melted the snow's crust and
made the going all the more difficult. The heat
from the sun melted the frozen clothing, making it
cling to their damp bodies. And the dazzling reflec-
tion of the bright sun off the snow blinded them.

In spite of the hardships and misery, they kept
on going. They expected soon to reach a cache of
food. It had been left a few days earlier by Captain
Tucker while he was on his way over the mountains
to rescue them. When they reached the tree where
the food had been hung, they were horrified to
find that wild animals had found it first.

Fortunately the new threat of starvation did not last long. The next day was the most wonderful day in the memories of the Reed family. At about noon they met Mr. Reed and a rescue party of 14. Mrs. Reed, exhausted, sank to her knees and said a prayer of thanksgiving. Virginia and Jimmy ran to their father. He threw his arms around them and there were hugs and kisses. Reed helped his wife to her feet, and for a long time he held her tightly as they all cried.

When Reed learned that two of his children were still at the cabins, he hurried on. If only he could reach them before they died of starvation. He flew over the snow, covering in two days the distance that it had taken five days for Virginia and the others to travel.

When he arrived at Donner Lake on March 1, he was overjoyed to find Patty and Tommy alive. However, the deathlike look of the famished little children made his heart ache. He made soup for the children and the other suffering survivors. He filled Patty's apron with biscuits. She carried them around, giving one to each person.

Reed organized a party of 17 which he would lead across the mountains. Three of his men were left behind to bring in wood and help those unable to travel. Reed also left behind seven days' provisions—not much, but far more than they had eaten in recent weeks.

Reed's party had not traveled far when a new storm broke upon them. The men worked all the first night trying to erect some crude shelters. For three days and three nights they were exposed to the fury of the hurricane of snow. As the storm continued, they ran out of food. When the storm let up, once again they were on the trail. It was a long hard trip, but they finally reached Sutter's Fort.

At long last the Reed family was all together again. Seated around a roaring fire at Sutter's Fort, Mr. and Mrs. Reed, Virginia, Patty, Jimmy and Tommy ate and made plans for a new life in California. The Reeds, the family who had suffered the most hardship on the first part of the trip, were all safe. They were the only family not to lose one or more members in the tragedy at Donner Pass.

CHAPTER FOUR

SNOWBOUND
ON A
TRAIN

CLICKETY, CLACK. Clickety, clack. Clickety, clack.

The train, *The City of San Francisco*, raced across the desert lands of Nevada toward California. She was carrying 196 passengers and a crew of 30. Outside it was cold and dark, and the winds were blowing a trace of snow. Inside the train it was warm and most of the passengers were asleep.

As the night wore on, the Southern Pacific's crack streamliner continued to make good time. After a short stop in Reno, the train started up the eastern slope of the rugged Sierra Nevada mountains.

Clickety. . . . . clack. Clickety . . . . .clack. Clickety. . . . . .clack.

The rhythm of the sound of steel wheels against the rails slowed when the train began its climb. As the train wound its way up the mountains, the wind increased, and it snowed harder and harder. By the time the train approached Donner Pass, the wind was gusting at 100 miles per hour, and the air was full of blowing snow. The engineer and firemen riding in the forward locomotive could barely make out the tracks. It was only when they passed through a snow shed or tunnel that they had any relief from the storm.

Up ahead a mile or so, another engine, pushing a rotary snowplow, labored to clear the tracks for the streamliner that followed. Suddenly an avalanche of rocks and snow came crashing down the mountain. The work train was quickly buried in a mountain of snow.

The high winds soon piled new drifts of snow across the tracks which had been cleared. The *City of San Francisco's* engines lacked the power to push her V-shaped snowplow through the huge mounds of snow. The train was forced to stop. All morning the crew jockeyed the train back and forth, trying to buck through the growing snowdrifts. At noon they gave up and settled back to wait for help.

That Sunday morning, January 13, 1952, was the beginning of what was to be a long wait. However, neither the passengers nor the crew was very concerned. The unexpected stop gave the trip a little extra excitement. Some of the passengers had never seen a real blizzard before. The conductor went through the cars telling the passengers,

"Everything's all right, folks. A plow is coming up the other way to get us out."

Actually the train was in more danger than anyone realized. It was trapped on a high ledge with a deep canyon on one side. On the other side was a steep upward slope. This slope was covered with a thick layer of snow which could come roaring down at any minute. If this happened, the stranded train would be buried, or worse yet, pushed off its scary perch. Ahead and behind the train, ever-growing snowdrifts would prevent any quick rescue.

Among the passengers were Dr. and Mrs. Walter Roehll of Middletown, Ohio. Back home Dr. Roehll had just finished a long and tiring year. Both he and his wife were looking forward to a week's vacation in the sun on the beaches of Hawaii.

As they had packed for the trip, Mrs. Roehll asked her husband if he planned to take along any drugs or medical equipment. He looked at the black bag and remembered how it had hardly been out of his sight during the past year.

"I think not," he said. "There will be other physicians wherever we go."

Then, for some reason, he changed his mind. He took from his medical bag 2,000,000 units of penicillin, a bottle of morphine, 50 half-grain tablets of phenobarbital, a bottle of aspirin and a hypodermic syringe. He placed them in his wife's pink, quilted hosiery box as she finished her packing.

Now he was glad that he had. There was a knock on the door of their train compartment.

"Are you a medical doctor?" a crewman asked.

Dr. Roehll nodded.

"Can you come with us? We have word that there is an injured man in the cab of the snowplow up ahead."

Dr. Roehll picked up his wife's hosiery box containing his supply of drugs and set off with several members of the crew into the storm. It was hard going through the howling wind and blowing snow. They had difficulty in seeing where they were going. The recently formed snowdrifts had only a thin crust, and their feet broke through with each step.

When they finally reached the snowplow, there was little they could do. The avalanche had overturned the snowplow before partially burying it. The man had been crushed by the toppled plow. The snowstorm had taken its first casualty.

After returning to the stranded passenger train, Dr. Roehll had little time to rest. The first order of business was to help the crew to find any nurses and other doctors who might be aboard. A half-dozen registered nurses were identified, but it turned out that Dr. Roehll was the only medical doctor.

Patrols were organized, and a nurse or someone who knew first aid was placed on duty in each car. In order to save water, the toilets in all but one car were shut down. When that one ran out of water, a toilet in the next car was opened and so on. The doctor also ordered smoking stopped in order to conserve the little fresh air that remained. Later he ordered the bar closed. "Alcohol and the proper attitude just don't go together," he explained.

Outside the worst snowstorm to strike the California mountains in 15 years continued. The space between the train and the high bank on one side soon filled with snow. Both air and light from the outside were cut off. A gang of track workers, who had been working nearby when the storm struck, struggled to keep the windows and doors on the other side clear of snow. If it had not been for their heroic work, the passengers would have soon suffocated.

Once things were organized aboard the train, Dr. Roehll returned to his compartment to rest. He had no more than sat down when there was a knock on the door.

"Come quick. My little girl—she's so hot."

Dr. Roehll, with his pink case of medicine, followed the worried father to one of the forward cars. The three-year-old girl had a temperature of 102.5°F. Her throat was raw and inflamed. The doctor gave her a shot of penicillin. He had to be careful to conserve the valuable medicine, so he could not give her as much as he would have liked.

A trainman touched his arm. "Doctor, there's a man in Car No. 5 that looks real bad."

The passenger was in great pain. Dr. Roehll asked a few questions about his medical history before giving him some morphine. As before, not much, but enough to stop most of the pain.

Dr. Roehll returned to his compartment. He dozed briefly only to be awakened again. This time it was a woman with an infection in the upper respiratory tract. Then a man with an earache needed his attention. Next he prescribed bed rest for the victim of a mild stroke.

As Sunday wore on, conditions aboard the train became more and more uncomfortable. First the air conditioning system broke down. It became difficult to breathe. There was a bad smell caused by too many people being crowded together.

Next the lights dimmed, flared up for a moment, dimmed again, and then went completely out. The big diesel engines which powered the train were out of fuel. Now there would be no more heat in the train. The few cars that had propane gas generators had dim lights. In the other cars it was totally dark.

The cars became clammy, dripping iceboxes. People put on one layer of clothes over another. They wrapped themselves in blankets, and a few tore down the window curtains to wrap around their feet. Friends and strangers alike huddled close together to keep warm.

As long as there was light, people's spirits seemed to be good. Once the lights went out, some worried aloud that a snowslide might send the train crashing down the mountainside. Others worried that they might run out of food or water, or freeze to death before help could reach them.

One passenger was familiar with the area in which they were snowbound. He estimated that they were only a short distance west of the Donner Pass, the place where the Donner Party had been

Deep freeze on wheels! Streamliner
City of San Francisco *snowbound near*
Donner Pass in 1952.

trapped by a snowstorm in 1846. The passengers were not encouraged by his telling that only 45 of the 81 people in the Donner Party had survived that famous snowstorm 106 years earlier.

Dr. Roehll and several others realized that the people's mood might be the most important factor in their struggle for survival. Flashlights and kerosene lamps were rounded up and placed in the cars that had no lights. Fortunately there was plenty of food aboard, and hot coffee and chocolate helped. Some of the passengers organized a song-fest in the club car. They started the singing with *California, Here I Come.* Strangely enough, the favorite songs were *I've Been Working on the Railroad* and *I'm Dreaming of a White Christmas.*

Late Sunday night, Dr. Roehll took one last tour through the train before turning in. All the passengers knew him and felt comfort in his being aboard. As he went from car to car, he stopped to give a word of encouragement or to say something to calm a frightened passenger.

As he worked his way through the train, he kept thinking of the dangers that went with such a cold and damp environment. If they weren't rescued quickly, he was afraid many of those aboard might develop respiratory diseases, such as pneumonia. With only a dozen or so shots of penicillin left, things could get worse in a hurry.

Then Dr. Roehll had another thought. What if he got sick? With so many people depending upon him, he didn't dare let that happen. And what he needed most now was some sleep.

During the night Dr. Roehll managed to get a few cat naps between calls for help. In the morning a steward brought him and his wife hot coffee and two plates of beans and frankfurters. When he had finished his breakfast, Dr. Roehll once again worked his way through the train. The air in the cars was growing stale, and in some cases, it was almost unbreathable. He asked volunteers to open windows and doors from time to time in order to get fresh air into the cars. Even though it was almost unbearably cold in the cars, the people had to have air to breathe.

The first good news came on Monday night. A group of skiers from a lodge about 20 miles away reached the train. They brought food, but the news they brought was appreciated even more. Efforts were under way to rescue the stranded passengers and crew. Suddenly everybody felt better. After the skiers had collected telegrams to be sent to worried relatives, they set off in the stormy night to return to the lodge.

Less than an hour later, Nurse Anna Lundblom was at Dr. Roehll's door pleading for him to come quickly. He was stunned by what he saw when he entered the next car. The passengers were gasping for breath. Many were sprawled in their seats, clawing at their collars. One man stood up when Dr. Roehll entered. The man gulped once and then fell to the floor.

Dr. Roehll quickly noted the symptoms. What could be causing the problem? Food poisoning? High altitude? Heart attacks?

Suddenly he realized that it was none of these—it was *carbon monoxide!* He shouted to one of the trainmen, "Open all the windows and doors! We've got to get some fresh air in here!" And then in almost the same breath, he yelled. "Turn off those generators!"

The propane generators had been started to provide light in the cars. No one realized that the exhaust pipes had become clogged with snow. As a result, the deadly gas had seeped up through the floor, affecting the 60 passengers in two of the cars.

Some of the passengers were locked inside their compartments. Dozens of steel doors had to be forced open before these victims could be rescued.

The first thing to do was get the stricken passengers breathing fresh air again. In some cases it was necessary to give artificial respiration. Once the victims had regained consciousness, there wasn't much that could be done except to give them aspirin for their headaches. Dr. Roehll also ordered them to stay near an open door or window and told them to keep bundled up so they wouldn't catch cold.

Dr. Roehll and the nurses who helped him also suffered from breathing too much carbon monoxide. One of the nurses, Helen Geurtz, described what it was like.

"It was awful," she said. "I thought at first that the victims had suffered from heart attacks. And then, as I worked on the unconscious ones, I began to feel warm. I took off two coats, and I still was hot. My heart was beating at a terrible rate."

Nurse Geurtz went on to say, "Dr. Roehll was close to passing out a number of times, but I never saw him stop working. Finally, about two o'clock in the morning, I passed out. The next thing I knew, I woke up with a bad headache. It was about 6:00 A.M. Some of the others were still at work taking care of their patients."

Fortunately there were no major problems on Tuesday. Uncomfortable and tired as they were, the passengers and crew made it through the day and the long night that followed. They woke on Wednesday to find a cloudless sky. The sun was shining brightly on the huge drifts of snow—the snowstorm was finally over.

Overhead several airplanes and a couple of helicopters circled. Some dropped food and medical supplies, and others carried reporters and photographers. Word soon came that giant snowplows were busy clearing a nearby highway which crossed the railroad tracks about a half mile up ahead. This was to be their escape route.

By Wednesday noon the faithful track workers had made a narrow path from the train to the highway. Then a strange looking procession began. The passengers, many wrapped in curtains and towels to keep warm, struggled through the snow. A few carried children in their arms. Four passengers had to be carried out on stretchers.

Once they reached the highway, they were quickly placed in automobiles and taken five miles down the road to a ski lodge. The first order of business was to get warm. Next was to have a hot meal. From the ski lodge, it was only a short walk to a rescue train waiting to take them to Sacramento and on to San Francisco.

Friday found Dr. Roehll and his wife in the comfort of a warm San Francisco hotel room. His energy restored by twelve hours of sleep the night before, Dr. Roehll talked to reporters about his experiences aboard the snowbound train.

"Dr. Roehll, what advice can you give to people who are trapped in a snowstorm?" one reporter asked.

The doctor thought for a moment and replied, "The same advice I would give in the case of any disaster. The key word is *conservation*— conservation of energy, of food, of water, and yes, even of worry."

CHAPTER FIVE

THE GREAT BLIZZARD OF '49

WITH THE POWER and swiftness of a prize fighter, the blizzard struck a surprising first blow. The country had hardly recovered from celebrating the start of the new year when the storm hit on Sunday, January 2. And it was a complete surprise—not even the Weather Bureau saw it coming. Their forecast for the western part of the United States was for "increasing cloudiness and possible snow flurries."

Wyoming and Colorado were the first to be hit by the swirling mass of snow. In less than an hour, travel by plane or automobile was impossible. Trains were soon brought to a halt. The biggest and longest tie-up in western railroad history had begun.

The blizzard spread. The Dakotas and Montana were soon in the grips of the cold, white nightmare. The storm crossed the Rockies to blanket Utah, Idaho and Nevada. Nor were Nebraska, Kansas or Oklahoma spared. In a matter of a few hours, most of the western United States was caught up in one of the nation's worst snowstorms.

Many blizzards, like the famous one in New York in 1888, strike a devastating first blow. Then they move on in a day or so and things are back to normal in less than a week. Not so with the blizzard of 1949.

The 1949 storm continued for seven straight weeks, from January 2 to February 19. Although there were a few breathing spells, six major blizzards and several smaller ones were recorded. Paralyzing storms were in progress somewhere in the west during 25 of the 48 days.

Winds reached 80 miles per hour. Temperatures dropped as low as 50°F below zero. Because of the howling winds, no one was quite sure how much snow actually fell. But in some areas more than 40 inches fell during the first three days. In Nebraska and Wyoming, ice seven inches thick covered railroad tracks and highways before they were covered with a thick blanket of snow.

Eventually the first snowfall stopped. Fierce winds whipped up the loose snow and visibility was cut to a few hundred feet or less. Snowdrifts quickly grew to 20 or 30 feet high.

All over the west roads were closed, telephone wires were down, and trains were stalled. And, because the storm caught people by surprise, thousands of unsuspecting travelers found themselves suddenly stranded. Many were only a short distance from the shelter of a house, gas station or roadside restaurant. However, they might as well have been a thousand miles from help. No one can survive on foot for long in the cutting snow and killing wind of a western blizzard. Many who tried to go for help lost their way in the blinding storm and perished.

The frozen bodies of Philip Roman, his wife and their children were found in a field near Rockport, Colorado. They had left their snowbound car and tried to cut across the open prairie to their home less than a mile away. The rancher's body was huddled over that of his son. His wife's body sheltered that of their daughter. The parents had wrapped part of their clothing around the children in an effort to save them.

Not far away, in Rockport, 343 stranded travelers spent three days jammed together in a tavern. Few had beds to sleep in. All got short rations as food supplies ran low.

The 58 people stranded in nearby Lone Tree fared much better. In their group were the drivers of two Safeway grocery store trucks. When the storm continued into the second day, the drivers opened their cargoes to feed the hungry people.

Many travelers were not as fortunate as those stranded at Rockport and Lone Tree.

The snowstorm caught Andy Archuleta, his wife and their five-year-old daughter on a highway. They were less than a mile from their home near Hillsdale, in southeastern Wyoming, when their old Ford coupe stalled in a huge snowdrift. Unlike the Roman family across the border in Colorado, the Archuletas decided to stay in their car.

When the car stalled, Andy got out to check on the situation. With snow up to his waist, he dragged a fence post to the car for fuel. Then he pulled off a hubcap from one of the wheels. He took it inside the car and built a small fire in it.

The Archuletas huddled over the fire. For hours, as the storm howled outside, they worked hard to keep the small flame going, coughing from the dense smoke. But slowly the numbing cold robbed them of their strength. As they sat snuggled together with their arms around each other, the fire went out. The wind blew the fine snow through every crack in the car. When rescuers finally reached the Archuleta family's frozen coffin, every inch of space in the car was tightly packed full of snow.

Railroad travelers fared little better than those caught by the snowstorm on the highways. Trains were halted, one after another. When the first storm ended, there were six passenger trains in the railroad yards at Omaha. Eight had been stopped at Ogden, Utah, five at Salt Lake City and five at Cheyenne, Wyoming. Six were stalled between Sidney, Nebraska, and Cheyenne. Altogether, 50 trains had been caught in the storm.

The Union Pacific's *City of San Francisco* was roll-ing across the plains of Nebraska as the blizzard grew in force. When the train crept into Kimball, the drifts were ten feet high. The engineer and fireman could see ahead for only 30 yards or so. The wind was gusting to 60 miles per hour or more. The temperature was down to 5°F below zero and still dropping.

The decision was made to stay in Kimball until the storm let up. At first the train's 270 passengers were able to enjoy all the comforts they expected on a modern streamliner. Then the train's water sup-ply ran low, and steam heat could no longer be provided. A steam engine was brought in from nearby Sidney, and the cars were kept warm for another few hours. However, the steam engine froze up during the second night, leaving the cars as cold as a deep freezer.

There was no choice but to move the passengers and most of the crew from the train to the town's two overcrowded hotels. The staff from the *City of San Francisco*'s dining car took over the kitchen at the Wheat Growers Hotel. Food was donated by merchants and townspeople, and no one went hungry.

A place to sleep was another matter. Beds were so scarce that many people had to sleep on the floor, tables and couches. Since most of the hotel workers were snowbound in their homes, the passengers did most of the work. One served as the desk clerk; others made beds and helped keep the hotel reasonably clean.

After three days the storm ended. A rotary snowplow fought its way into town. Twelve hours later the *City of San Francisco* pulled out. The old timers in Kimball, Nebraska, are still talking about the few days in '49 when the town's population of 2,000 unexpectedly increased by several hundred.

Once the first storm died down a bit, thousands went to work to rescue the stranded motorists and rail travelers. Food and medicine were rushed to isolated villages and farms. Bales of hay were dropped from airplanes to livestock and wildlife.

Until the railroad tracks and highways could be cleared, airplanes were the best way to bring aid to the storm victims. Radio stations broadcast directions for signaling low-flying aircraft. A single line in the snow meant that a doctor was needed. Two parallel lines was a request for medical supplies. A large X signaled a stalled vehicle. An F was a plea for food and an L for fuel. "All is well" was signified by LL. A triangle in the snow indicated a place where it was possible for an airplane to land.

Airplanes belonging to the Air Force, the National Guard, the Civil Air Patrol, and to private groups and individuals joined the gigantic rescue effort. Many of those involved in the rescue operation suffered as much as the storm victims they were trying to aid. William Harrison of Granby, Colorado, flew relief missions until his small plane crashed into the side of a mountain. For six days he lay in the snow, his feet frozen, before other fliers spotted his distress signals and risked their own lives to save him.

As January wore on and the snowstorms continued, one on the tail of another, the situation
became worse. Although most of the people
stranded by the first storms had been rescued or
given aid, livestock in the western states were in
great danger. In Nebraska alone, over two and
one-half million cattle faced certain death unless
help came soon. Another million and a half cattle
and sheep in South Dakota and Wyoming were
helpless. Standing with their backs to the wind and
neck deep in snow, they would be lost to starvation
if they did not get food quickly.

The governors of the western states asked the
Federal Government in Washington for help.
Military officials already were doing everything
they could, but it was not enough. Loss of the livestock would mean later meat shortages and higher
prices for citizens in all parts of the country.

President Harry S. Truman put Major General
Lewis A. Pick in charge of the Fifth Army's *Operation Snowbound.* It was General Pick who had built
the famous Burma Road during World War II. In
1949 he was in charge of building over 100 dams as
part of the huge Missouri River flood project.

General Pick called each of the private contrac-
tors working on the dams. They all agreed to send
heavy bulldozers, road graders and snowplows to
help clear the snow-clogged highways. Although
airplanes could carry enough food and medicine
for the marooned people, it would take trucks to
haul enough hay to feed the millions of starving
cattle and sheep.

In about 40 days the Fifth Army blasted open over 100,000 miles of highway. Almost a quarter million people, along with four million head of livestock, were freed from their snowy and isolated retreats.

The battle against the blizzard continued in the air as well as on the ground. Colonel Joe McNay of the Tenth Air Force was in charge of operations at Offutt Air Base near Omaha.

"We've got about 80 airplanes on the job," he told a reporter. "Everything from C-82's to helicopters and RF-80 jets. During the past ten days we've dropped 1,500 tons of hay, 1,000 tons of flour and 36 tons of food rations. That's not to mention 20 tons of coal, 38 tons of spare parts for snowplows and bulldozers, 12 tons of alfalfa pellets, six tons of aircraft skis and six tons of fresh milk. We've also dropped 5,000 blankets, 775 gallons of fuel oil, several tons of baby food and 14 tons of warm clothing.

"We've dropped everything the people have asked for, whether it's to a small farm or ranch or a snowbound village. Why, the other day we even dropped a half ounce of radium that was worth $500,000. A hospital in western Nebraska needed it badly.

"We've also evacuated nearly 300 expectant mothers from villages and ranches. We've been keeping pretty busy," he added.

While *Operation Snowbound* went on in Nebraska, Wyoming and the Dakotas, the Sixth Army performed a similar mission farther west. Huge C-82

"Flying Boxcars" dropped tons and tons of baled hay to cattle on isolated ranches in Nevada and Utah as part of "Operation Haylift."

Operation Snowbound was not all successes. New storms and 70-mile-an-hour winds often closed the roads as fast as they could be opened. The battle went on until late February, when General Pick could finally report that the West had been saved from the most widespread and severe snowstorm that anyone could remember.

No one will ever know the total cost of the *Great White Death,* as one writer called the Blizzard of '49. The human death toll was well over a hundred. A livestock census in Wyoming by the U. S. Department of Agriculture showed losses of 125,000 sheep and 23,000 cattle. Losses of livestock in Nebraska were believed to be even greater. Perhaps more than a million head of sheep and cattle perished in all.

The great winter storm that began on January 2 left its mark on the other parts of the country, as well as the ranch country. A series of tornadoes broke out in the warm air ahead of the cold air across Arkansas and Louisiana—two months before the beginning of the tornado season.

Nor did sunny Southern California escape the horrors of the winter of '49. Los Angeles had a record low temperature of 27°F, and the resort town of Palm Springs an icy 22°F. Oranges and vegetables in the Imperial Valley and other parts of Southern California were badly damaged, with losses estimated in the millions of dollars.

With spring in 1949 came a new fear. With record amounts of snow on the ground, severe floods seemed to be a sure thing. But the weather cooperated—warm enough in the daytime to slowly melt the snow and cool enough at night to freeze it. As a result of the slow melting, much of the snow water was absorbed by the soil and little flooding occurred. At least the western farmers and ranchers benefited from the Great Blizzard of '49 by having plenty of water for that summer's crops.

CHAPTER SIX

WINTER
AT
VALLEY
FORGE

THE AMERICAN REVOLUTION lasted through many winters—from March, 1770, when the first Americans were killed by British troops, until December, 1783, when George Washington was finally able to resign his job as Commander-in-Chief of the American Army. The snow, ice and freezing weather of these winters sometimes helped the Americans and sometimes caused them to suffer terrible defeats.

The first real winter battle of the Revolution was fought in Canada. There were two separate American forces invading Canada during the fall and winter months of 1775. The larger force was led by

General Richard Montgomery, who managed to capture a number of British forts and drive the English soldiers from the city of Montreal. A smaller force of Americans, led by Colonel Benedict Arnold, left Cambridge, Massachusetts, in a driving, freezing rain. For 45 days the soldiers stumbled through the woods. Poorly equipped and clothed, many of Arnold's men died during the heavy snowstorms in early November.

What was left of Arnold's army finally reached the banks of the St. Lawrence River. There, on the opposite bank, towered the fortress of Quebec. The Americans rested until they were joined by General Montgomery's troops and about 800 Canadian volunteers.

Montgomery waited to attack until he had the kind of weather he wanted. December 26 was a clear, very cold day. The British inside the fort felt safe from attack since it was too cold for a man to handle a musket.

As the mass of cold, dry Arctic air slowly moved away, the temperature began to rise. On the 29th the weather was clear and the British prepared for the attack. But General Montgomery waited. He wanted snow to cover his advance.

At dusk on December 30, the snow began to fall. The night was dark, and the snow fell harder and harder. Finally at four o'clock in the morning, Montgomery gave the word. Signal fires were lit, but in spite of the heavy snow, the British saw them. A terrible hail of bullets from above smashed into the Americans as they tried to climb the icy hill.

The paths along the river were blocked in dozens of places by river ice that had drifted ashore. Everyone soon realized that taking the fortress was impossible. The Americans withdrew to spend a horrible winter camped in the snow alongside the river.

Meanwhile, Washington's troops still surrounded Boston. There seemed to be no way to get the British army out of the city. The Americans now had some heavy cannon, howitzers and mortars which they had captured at Ticonderoga in May. But, 260 miles of muddy, primitive trails and many rivers separated the heavy guns from the siege at Boston.

The man in charge of the artillery was Henry Knox. A bookseller by trade, Knox knew little about guns, but he had a brilliant idea. He thought that he knew how to use the cold weather to his advantage.

As the weather grew colder and colder, Knox hurriedly had 42 huge sleds built. He also bought 80 yoke (pair) of oxen. As Benedict Arnold and his men in Canada swore at the bitter cold, Henry Knox prayed for colder temperatures and more snow. Finally, on December 21, a strong wind from the northwest announced the arrival of a mass of cold, Arctic air and snow. Knox loaded his guns onto the sleds and began to move toward Boston.

The heavy weapons made good time, once they started. Then, in the middle of the same violent snowstorm that frustrated the Americans' attempts to take Quebec, Knox reached the Hudson River at

Albany, New York. To his surprise and dismay, he found the river still unfrozen.

There he was stalled until January 10, 1776, when a hard freeze set in. The river froze. The heavy snow covered the trails. Instead of impassable mud, the heavy guns slid easily over the frozen ground. On January 25, the long line of sleds reached the hills around Boston.

The delay had cost Washington a great deal. All of the American soldiers were volunteers whose enlistments were up on the last day of December. Washington spent most of his time during the last days of the month talking to his troops, begging them to stay. "Knox will be here," he swore. "The guns we need are on the way." Some of the men did stay.

The arrival of the guns did a great deal to encourage the men in the cold camps around Boston. They went to work immediately, digging emplacements for the guns. Working only at night, they managed to have most of the weapons in place before the British understood what was being done. Faced with a constant bombardment, the British called for their ships and abandoned Boston in March. The winter of 1775–76 had given the Americans one terrible defeat at Quebec and one great victory at Boston.

Perhaps the most important winter of the war was during the last months of 1777 and the early months of 1778. Many members of the Continental Congress were beginning to doubt George Washington's ability to lead the army. Many other people were certain that the Americans were going to lose the war.

The summer of 1777 had been a bad one for Washington. The British had left New York City and had landed on the northern shore of Chesapeake Bay in Maryland. The Redcoats had then moved toward Philadelphia, where the Continental Congress was meeting. Washington had met them on September 11 and was badly beaten, losing 700 men. Philadelphia fell to the enemy, and the members of the Congress had been forced to flee for their lives. Washington then attacked the British main camp at Germantown in early October. Caught by surprise, the British army fell back. Then the Redcoats regrouped and counterattacked. Washington found his army retreating again.

As winter approached and the English troops settled down in the comfort of Philadelphia, many Americans expected Washington to attack them there. Other people expected him to take his army away from the threat of attack and to spend the winter far to the south. But Washington made neither of these moves. To everyone's surprise he moved to a winter camp in a group of hills not more than 20 miles away from the enemy in Philadelphia. The little village near which Washington spent the winter of 1777–78 was named Valley Forge.

December 17, 1777, found the ragged American army making its way slowly down the banks of the Schuylkill River. They splashed through the icy water and into the tiny cluster of houses called

Gulph Mills. There they made a temporary camp in a heavy snowstorm. Their baggage train had been sent along a safer route, and the 11,000 tired men slept in the snow without tents or blankets.

"I could have wept tears of blood," one soldier wrote the next morning. "I am sick. There's nothing to eat. Nothing for the animals either."

All the next day they struggled through the snow. They moved ahead barely one mile each hour along Gulph Road. Wolves howled in the woods along the rutted, frozen road. Thousands of the men were without shoes, blankets or coats. The hard ridges of frozen mud cut through the soles of their bare feet. General Washington, riding along at the rear of his troops, followed a bloody trail.

Washington's army reached Valley Forge on December 19. Their food supply was almost gone. Thanksgiving Day had been celebrated with half a cup of rice and a tablespoon of vinegar per man. Many of the American soldiers had already deserted, and most of the 11,000 who did go to Valley Forge swore that they would leave as soon as their enlistments were up at the end of the year.

Work began immediately on fortifications on a high ridge facing south and on cabins for the soldiers. Washington's aides had taken over a large house for his use. But the General told his troops that he would live in a tent until every man in camp was under a solid roof.

In spite of the snow and temperatures near zero, thousands of trees were cut, and nearly a thousand cabins were finally built.

Twelve men lived in each of these tiny buildings. They cooked over the open fire and slept wrapped in blankets. Their beds were piles of straw thrown on the muddy ground, or bunks made from the trunks of small trees.

Two thousand men gave up and went home at the end of December. Fifty of Washington's officers resigned and left the camp. There were rumors that Washington himself was about to resign or be fired. Supplies of food and clothing finally stopped coming as the winter wore on.

A cold wind swept across the broad valley that lay to the south of the camp. Eleven hospitals had been built but, by Christmas Day, these were overcrowded. The sick and dying soldiers were then taken to barns, churches or any other buildings that could be used.

No one knows exactly how many men died in Valley Forge that winter. The number was probably close to 3,000. The bodies of the dead were stripped of everything that could be used by someone else. The naked bodies were buried at night in unmarked graves. This was done to try to keep the British spies in the camp from finding out exactly how weak the army was becoming.

In spite of the problems in the camp, Washington's choice had been a good one. Here at Valley Forge he was able to keep close watch on the enemy in Philadelphia. His camp was protected on the north by the large Schuylkill River and on the west by a deep gorge that had been cut by a stream known as Valley Creek. And an easy escape route

*The American
troops at
Valley Forge
trying to get warm
at a campfire.*

*General Washington
and Lafayette
visiting the cold
and hungry troops
at Valley Forge.*

into the rugged, wooded mountains of Western Pennsylvania lay nearby. The valley to the south and east was protected by gun emplacements that stood on a high ridge. A second line of cannon stood on a high, wooded hill in the middle of the camp, ready to slaughter any army that tried to cross the open fields below.

The camp was secure, but everyone in it was cold, hungry and sick. On January 25, there were only 25 barrels of flour in the camp. One-half of the men reported for sick call that day. Those that were able to stand watch had to stand on their hats in order to keep their feet from freezing.

"What did you have for breakfast, soldier?" someone would cry out.

"Fire-cake and water, Sir!" a hundred voices would reply. Fire-cake was a mixture of flour and water, fried on a rock placed in an open fire. It was a sticky paste, often full of soot and dirt.

"What did you have for dinner, soldier?" someone else would shout.

"Fire-cake and water, Sir!" would always be the answer.

"What did you have for supper, soldier?"

"Fire-cake and water, Sir!"

Fortunately for the new country, the winter of 1777–78 was only moderately bad. If they had made camp at Valley Forge two years later, the entire army would have died. As it was, the ground was covered with snow through all of January, but during less than half of February. By early March, the coldest part of the winter was over.

And, suddenly, so was the famine. During the last few days of February, thousands of pounds of food unexpectedly appeared right in the middle of the starving camp!

A bridge had been built across the Schuylkill River, and a group of soldiers were always camped nearby ready to defend the bridge against a British attack. One of the guards noticed a strange motion in the water. It seemed to tumble and roll, and the sunlight flashed from it in a peculiar way. The curious soldier crossed to the center of the bridge and looked down into the shallow water. The river, from which everything alive had been caught and eaten weeks before, was now suddenly alive with fish!

The fish were shad, and they had come to the camp at Valley Forge all the way from the Atlantic Ocean. Each spring, these large fish swim up certain rivers to lay their eggs. But they usually do not move as far inland as they did in 1778.

"Fish! Fish! The river's full of fish!"

The welcome cry flew across the camp. Officers and men tumbled from their cabins. Orders were shouted. A hundred mounted cavalrymen formed a line across the river and drove the fish toward the camp. Hundreds of other soldiers netted the fish. The shad run lasted for two weeks, and when it was over, Washington's army had enough salted fish stored away to last them until late summer.

But, even though the winter had not been as bad as it might have been, it still had been a terrible time for the American army. Over 1,500 valuable horses had died of starvation. At one time, Washington had only 5,000 men in his camp and between one-fourth and one-half were ill. Smallpox and typhus were everywhere.

The British spent the winter of 1777–78 comfortably in the houses of Philadelphia. They planned to let the "ragged band of rabble" suffer as long as possible and then, when they had become weak and defenseless, defeat them once and for all.

"The Revolution is dying of natural causes," the British Commander wrote to his King. "This will be the last winter of the war."

The British had good reason to believe that they would soon win the war. The American rebels had proved to be weak soldiers. They could not march well. Their knowledge of how to fight with bayonets was so poor that the weapons were used mostly for holding meat over cooking fires. The American army had been unable to win any battles against the British army or their mercenary German allies unless they were favored by the weather, or by the element of surprise, or by having more men.

But the British did not know that the men in Washington's camp at Valley Forge were busy. The half-starved, ragged, cold, sick men were spending every day learning to be soldiers. The British were in for a surprise when the spring battles began.

The man who taught the Americans how to fight was a 47-year old German, Baron Friedrich Wilhelm von Steuben. In many ways, he was one of the most interesting men to have fought in the

American Revolution. For the most part, he was a fake and a fraud. He passed himself off as a Lieutenant General, but the highest rank he had ever held was Captain. None of the armies of Europe would give von Steuben a job, yet he claimed to be an expert in military matters. But the Americans needed all the help they could get. So, Washington accepted the German's offer to serve.

Von Steuben's reputation may have been faked, but he did understand the German method of military drill. He quickly realized that each of the detachments in Washington's army had different regulations. This made it difficult for them to work together as an army. He immediately started writing a drill manual for the Americans. Unfortunately, the Baron did not know English. He wrote out his instructions in French. His secretary translated these into a rather formal form of English. Two American officers then had to rewrite the manual into the type of English the American officers and men would understand.

Von Steuben himself directed many of the drills, his orders being translated by an aide. Practice with the bayonet was one of his favorite drills. He also taught the Americans how to load their muskets more quickly and how to march efficiently.

As the trees turned green and the flowers bloomed in the fields, the British army began to move. Washington left Valley Forge at the head of an army of professional soldiers. The men were still ragged and barefoot, but they marched smoothly and quickly in long, straight columns. Each musket carried a bayonet that gleamed cleanly in the sun. The men sang *Yankee Doodle* as they marched proudly along to the music of the drum and fife.

On June 28, 1778, the British army was attacked near Monmouth Court House, New Jersey. The first attack, led by Major General Charles Lee, went badly, and the American troops fell back. Von Steuben was sent to reorganize the retreating men. Soon they returned to the battle, marching in perfect order. With slashing bayonets, they cut through the British army. This time it was the Redcoats who retreated.

The war was far from over, but the tide had turned. The new country had managed to survive its darkest winter.

CHAPTER SEVEN

THE
BLIZZARDS
OF
1966

THE UNITED STATES NEWSPAPERS PUBLISHED during the first three weeks of January, 1966, didn't carry much exciting news. The war in Vietnam was still going on, although President Lyndon B. Johnson had called off the American bombing raids for a while. Congress was busy trying to decide whether they should provide more money for the war. A U.S. Air Force transport plane had crashed, killing 42 infantrymen of the First Cavalry Division.

The new James Bond movie *Thunder Ball* had been out for only seven weeks but was setting attendance records everywhere. On the basketball court, Wilt "The Stilt" Chamberlin scored 53 points as the *76ers* beat the *Lakers* in New York's Madison Square Garden.

Even the weather map seemed unexciting. There was a huge high pressure area in Canada, just north of the Great Lakes. Cold, Arctic air spun out of this high in a clockwise direction. As it did, it picked up a little moisture and threatened New England and the Great Lakes area with snow flurries. A low pressure area in the Gulf of Mexico pulled the air from the high toward the south, bringing snow showers to many of the coastal states. It was chilly in New York City with temperatures in the twenties.

The most exciting weather news wasn't reported in the newspapers because few people realized how important it was. More than six miles up, high above the United States, a strong wind blew from west to east. Known as the *jet stream,* this hundred-mile-an-hour wind normally crosses the East Coast near Philadelphia. During the winter months the jet stream usually acts as a block between the warm, wet air from the south and the cold, dry air from the Arctic. But during January, 1966, the stream of air suddenly shifted southward. The low from the Gulf slipped quickly up the coast under the jet stream which had become southwesterly over that section. On the 26th, the low pressure was centered off the coast of North Carolina.

Cold air flowed down across the Great Lakes, through the mountains of the east and into the deep south. The temperatures dropped to 35°F in semi-tropical Miami, Florida. Ducks became trapped in freezing ponds and lakes in Alabama as the temperature reached a record low of 24°F *below zero!* The steadily moving water of the mighty Mississippi River began to freeze, and ice jams blocked river traffic. Reports of record snowfalls began to come in from Kentucky, Tennessee and the Carolinas. A blanket of snow, three feet deep, lay on the mountains of the Blue Ridge in Virginia. Winds of 50 to 60 miles per hour whipped the snow into huge drifts. One drift was so deep it reached the second-story windows of President Johnson's home, the White House.

The snow began to fall in New York City at 2:30 in the morning on January 27. Six inches were predicted for the city. Remembering the terrible blizzards of the past, city officials quickly made plans. Nearly 2,000 men were called to work, and 1,200 pieces of snow-moving equipment were made ready. The temperature dropped to the teens.

Oswego, New York, is a port city on the southeast edge of Lake Ontario. The weather men there watched the developing storm carefully. As the low from the south moved off the coast of New England, a second low pressure area suddenly appeared on their maps. This one lay over Canada, just to the north of the Great Lakes. Temperatures in Minnesota fell to 31°F below zero. Chicago reported a temperature of only one degree above zero, with a wind of 20 miles an hour blowing in across Lake Michigan.

Snow began falling in Oswego a little before noon on Thursday, the 27th. Snow showers continued through the rest of the day, and by nightfall only about one-half inch of tiny snow crystals covered the ground. Then, at a few minutes after 5 o'clock, the situation suddenly changed. The wind grew stronger and stronger, finally reaching gusts of 40 miles an hour. The snow fell harder and harder. By midnight more than eight inches had fallen.

The next day both of the lows had moved off the coast of New York. A violent wind circled into them in a counterclockwise direction. On the west side of the lows, the wind blew from north to south, pouring frigid air into New York State.

The temperature in New York City fell to 8°F. Many of the families that lived in the city's slums were without heat. The National Guard Armories and five YMCA buildings were quickly turned into dormitories, and the Salvation Army set up free food kitchens. Fires broke out all over the city as heating systems became overloaded. On this one weekend, 11 children were to die in these fires.

In Oswego, the storm grew worse. The wind still blew at more than 25 miles per hour, and the snow fell steadily. Visibility was no more than a few feet in the swirling white world. As the drifts became deeper, traffic slowed and finally stopped. By noon, a foot of new snow lay on top of the eight inches that had fallen the night before. Suddenly, shortly after noon, the wind slackened and shifted direction. The snowfall lessened and the clouds seemed to lighten a little. The people of Oswego thought that the storm was over. They began to clear the snow from their driveways and sidewalks.

Very little snow fell anywhere in the United States during Saturday morning. But it was terribly cold. It was almost as if a new Ice Age had struck the eastern half of the country. Bismark, North Dakota, reported that 10°F below zero was the *warmest* temperature they had that day. Des Moines, Iowa, and Chicago, Illinois, both had high temperatures of −6°F. The temperature got down to −10°F in Amarillo, Texas. Way down south in Mississippi, the temperature ranged between 14°F and 20°F all day. The thermometers in Kansas City stayed at 8°F, and in Richmond, Virginia, it never got above 18°F.

New York State was frozen solid. Temperatures in the teens were reported from all over the state. Snow began to fall again in Oswego around noon and in New York City shortly after dark.

When people on the East Coast woke up the next morning, they were glad it was Sunday. No one wanted to go out-of-doors that day. Two convicts who were due to be released from the prison in Rochester, New York, decided to stay in jail until the storm let up.

Seven more inches of snow fell in New York City. To the west, the fall was even heavier. Oswego got another 21 inches, making a total of 52 inches on the level ground. With winds that often reached 50 miles per hour, drifts began to build. Every airport north of Charleston, South Carolina, was closed, with only a few exceptions. Every highway in New York State was closed. The few trains that could run were often four or five hours late.

The low pressure area moved off the coast of New York, just as the one before had. By Monday morning the low was pumping frigid air into the state from the Arctic. Snow continued to fall steadily in Oswego, as well as in most of the eastern part of the country. Another 50 inches of new snow had fallen on Oswego before midnight, when the storm finally stopped.

Oswego now had nearly 102 inches of snow on the ground. Stop for a moment and think about how much 102 inches of snow really is. That is eight and a half feet! Can you imagine snow lying so deep on a basketball court that it is within 18 inches of the basketball hoop?

The area covered by this storm was 1,500 miles long and 500 miles wide—750,000 square miles. More than 35 million people lived in this region of the United States. Of these, more than 200 died as a result of the blizzard.

Slowly the country returned to normal. The airports opened, and the trains returned to their schedules. Snowplows moved drifts from the highways, and automobile travel resumed. The war in Vietnam went on, and Congress voted to provide the military effort with $4.8 billion. The Russians put a spacecraft on Venus, but it sent back very little information. The *76ers* won another basketball game.

The weather map for March 1 showed three low pressure areas. One was off the East Coast, one over Nevada, and the third near Oklahoma City, where the temperature was a comfortable 66°F. By the second of March, the two lows in the center of the country had joined to form one huge depression, now lying over central Kansas. By the morning of the 3rd, the low had moved into Iowa. To the south of this whirling mass of air, the temperatures were still comfortable. But Bismark, North Dakota, reported temperatures of between 13°F and 17°F.

Snow was falling in the northern Great Plains. Blizzard warnings had been broadcast throughout the region. With sighs of disgust, the people of the Dakotas got ready for another day or two of snow. Cattle were moved in close to the houses. Food for both the animals and the human beings of each community was carefully stored away. Travel plans

A Thruway Junk Pile? Automobiles shoved aside by snowplows fighting to open the superhighway after the storm of 1966.

were cancelled. The winds grew to gusts of 30 miles per hour, and the road-clearing crews were kept busy.

The low pressure area that caused the storm moved quickly toward the northeast. Then it ran into a huge high pressure area that lay over southern Canada. The snowstorm stalled right over the Dakotas. The wind howled at hurricane force, driving the fine snow through cracks around doors and windows. From two o'clock in the morning of March 3 until seven p.m. on the following evening, the visibility was down to less than one-eighth of a mile. During at least ten of those hours, no one could see anything at all.

A snow-blinded calf stands stunned and lost during one of the worst storms in recorded history in February, 1966.

Near Strasburg, North Dakota, a six-year-old girl became lost in the swirling snow as she tried to go with her brothers to their barn. In a walk of less than 60 feet, the boys lost sight of the little girl. Her body was found later only a quarter of a mile away.

In nearby Woodworth, a 12-year-old girl heard the wind banging the door of the chicken house. Knowing that the chickens would freeze, the girl tried to walk the 100 feet that separated the house from the chicken house. But she never returned home. The next day, she was found frozen to death.

More snow fell as the storm began to move again. In Minnesota, the wind reached nearly 100 miles per hour at times. The wind whipped the snow into deep drifts. The Northern Pacific's train, *The Mainstreeter*, left New Salem, North Dakota, bound for the West Coast. But only one mile out of town, the train's three diesel engines plowed into a snow drift more than 20 feet high. The 100 people on board had to be rescued from the snowbound train and taken back to town.

Many other people were also snowbound. Near Mandan, North Dakota, three basketball coaches were trapped in their car for nearly four days. As the temperature fell well below zero, they built a fire inside the car. For food they had only three sweet rolls. Nearby, a farm family ate 15 candy bars as they waited for help in their car buried in deep snow only two miles from their home.

Not everyone was as lucky as these people. Two women died while trying to walk home from their stalled cars. Others died from heart attacks, asphyxiation and fires. In all, the blizzard claimed the lives of 18 people and $12 million worth of livestock.

But winter was not yet over. The last blizzard of the season struck Iowa and surrounding states on March 21. The temperature was well above freezing as a low pressure area drifted in from the west. Lightning bolts jumped through the sky, and a springlike rain fell. As thunder rolled back and forth between towering clouds, many people felt that winter was finally finished. But this thunderstorm was the beginning of a blizzard that would kill 27 people.

Most low pressure areas that develop in the far north form over the land. Because of this, they are normally dry and cold. Lows such as these often bring cold temperatures to the northern states. Usually, very little rain or snow will develop from them. They move rapidly eastward so that Gulf moisture does not have time to get drawn into them.

But the low pressure area that moved into Iowa late in March, 1966, had formed over the ocean, just off the coast of Alaska. It was not only very, very cold, but it was also very, very wet. The warm rain quickly changed to sleet as the cold air mass moved toward the east. Then the wet, clinging snow began to fall. By nightfall on the 22nd, 8½ inches of snow lay on the ground around Sioux City, Iowa. Farther to the northeast, the storm dumped 11 inches of snow on the Minneapolis-St. Paul area in less than 12 hours. Behind the snow came winds of up to 70 miles per hour which whipped the snow into drifts. In some places banks of snow 15 feet high blocked roads and covered stalled cars. Schools closed and businesses shut down to wait out the storm.

A fall of 8½ to 11 inches of snow is not a lot for this part of the Middle West. It was the wetness of the snow that made this storm so deadly. Normally, a snowfall of 10 to 12 inches is equal to one inch of rainfall. But the snow in the late-March storm of 1966 contained nearly twice this much moisture.

The wet snow clung to everything that it touched. Trees that had stood up under the weight of hundreds of heavy snows bent and broke. Power and telephone lines became too heavy and snapped. Tens of thousands of telephone poles broke under the weight of the snow and the pressure of the winds. From the eaves of houses, icicles were found sticking straight out parallel to the ground.

Trapped in their homes by the high snowdrifts, the people of Nebraska, Iowa, Minnesota and Michigan found themselves without electric power. Their furnaces stopped, and their homes became uncomfortably cold. At least 15 people died in their snowbound cars, and another 12 were frozen to death as they tried to walk through the howling wind and snow.

Unlike the storm earlier in the month, there was no high pressure area blocking the route to the east. By March 25, the storm's center had moved out over the Great Lakes. Spring was finally on its way.

CHAPTER EIGHT

AVALANCHE IN THE LAND OF THE INCAS

"COME! It is time to go!"

The short, barrel-chested man hitched his brightly colored poncho around his shoulders as he shouted to his sons. The three boys were dressed exactly like their father. They wore knee-length trousers, loose shirts, black leather sandals and wool ponchos. All of them had black hair and skin the color of light chocolate milk.

The boys shouted and whistled until the scattered sheep were gathered into one woolly mass and headed toward home. Their father followed a few yards behind, leading his prize possession—a huge llama. The sheep were the property of the

rich man who owned this land, but the llama belonged to the shepherd. He talked to the animal
soothingly as they walked across the slanting pas-
ture land.

The language these people spoke to their animals
and to each other would sound very strange to us.
It is *Quechua,* an ancient language used in these
mountains of Peru for many hundreds of years. It
is the language of the ancient Incas.

The shepherd had often talked with his sons
about their ancestors. He told them stories of the
Inca Empire that began about 800 years ago on the
banks of Lake Titicaca, a few miles south. The
shepherd knew the legends well. He had learned
them from his father. It was important for a man to
know that he belonged to a great race of people,
especially when he was now little more than a slave.

The Inca tribe began, the shepherd believed,
when the son and daughter of the sun rose from
the water of Lake Titicaca. They had been sent to
the earth to teach men how to live together under
laws, how to use the land to grow more food and
how to worship the sun. To do this, of course, it was
first necessary for the Incas to conquer their
neighbors. Generation after generation of wars
carried the Inca civilization farther and farther
away from the shores of Lake Titicaca. By the year
1400 A.D., the Inca Empire controlled the Andes
mountains from the equator to a point nearly 5,000
miles to the south.

A complicated form of government was developed during this time. Cities grew, and 10,000

miles of roads were built through the mountains to connect them. Irrigation was developed, and an abundance of food was grown and distributed throughout the Empire. Gold and silver were taken from the earth. The Inca Empire became both rich and powerful. The shepherd and his sons were proud of their Inca heritage. Most of the people in the valley called them "Indians," but they thought of themselves as Incas.

The shepherds reached their home. The boys herded the sheep into a large pen where they would spend the night. The father led the llama through the doorway and into the small adobe house. Such a fine animal as this was too valuable to be left out-of-doors and would spend the night with the family in the single room.

The women of the family were digging a few potatoes from the family's garden. The father called to them from the house. He was going to walk to the nearby village of Pacucco, he said, and would return shortly after dark. His family could not know that this would be the last time they would see him.

The day had been clear and warm. The sun was now rapidly dropping toward the Black Mountains, across the valley to the west. Long shadows crept toward the Santa River, shining far below him. He would leave Pacucco as soon as the lights went on in Ranrahirca, he promised himself.

Ranrahirca was the largest of perhaps a dozen small villages and towns that dotted the narrow floor of the valley.

The shepherd did not like the towns of the valley and rarely went lower into the valley than the village of Pacucco which perched in a canyon high in the foothills. Here the houses and shops were made of adobe and had dirt floors, just like the houses of the Incas who worked the mountain slope. In the towns below, the houses were made of stone that had been whitewashed. The streets were paved with cobblestones. Every night, as soon as darkness fell, the town was brightly lighted with electric lights. Few Incas felt comfortable in such a place.

Most of the people who lived in the towns of the valley could trace their ancestry back to Spain. They were as proud of their background as anyone in Peru. After all, it was the Spanish who had conquered the mighty Inca Empire.

The first Spanish explorer landed in Peru in the year 1522. Eight years later, a Spanish army attacked the Inca Empire with superior weapons, horses and a knowledge of warfare unknown to the Incas. The Inca civilization collapsed quickly, and by 1533, Inca gold and silver were flowing to Spain. The proud Inca Indians became slaves and were forced to work in the mines and in the fields.

The Spanish controlled Peru and much of the rest of South America for more than 300 years. A series of civil wars finally freed Peru from Spain in 1826 (although Spain did not recognize Peru's independence until 1879).

But freedom from Spain did little to change the social structure of Peru. The people who own the land today are the descendents of the Spanish conquerors. The businesses and the government of Peru are also in the hands of people of Spanish ancestry. These people live in large, modern cities along the Pacific coastline, or in smaller villages scattered throughout the valleys of the Andes Mountains.

Ranrahirca was a typical mountain village. The people who lived there spoke mostly Spanish and lived much like other small-town people throughout the world. On that Wednesday evening in early January, 1962, normal, typical things were happening. A business man, just returned from a trip to Lima, rested in the quiet of his comfortable home. Across the street, a family party was just getting under way. Across town, a group of children shouted and laughed as they played games in honor of the birthday of a friend. The town's electrician excused himself, leaving his friends in order to throw the switches that would give Ranrahirca electric power throughout the night.

The mayor of Ranrahirca was out for his usual evening stroll. He was proud of his town and liked to look it over carefully just as the sun was setting. He walked through neat, clean, cobblestoned streets. A pair of new houses were under construction, a sign that more important people were moving here from the crowded cities along the coast.

"This should be one of the major tourist attractions in South America," the mayor told himself as he walked along. "I wonder what we should do to attract more tourist business. We have everything here. People have always liked this valley."

The valley in which Ranrahirca lay was called the *Callejón de Huailas*—the *Corridor of Greenery.* Some of the tourist books called this area *The Switzerland of Peru,* but its mountains dwarf any of the famous peaks of Switzerland. As a matter of fact, the Andes are taller than any mountains in the world except for a few of the peaks in the Himalayan Range of Asia.

To the west of where the mayor walked were the Black Mountains. For the first time in the memory of anyone in Ranrahirca, a small amount of snow had fallen onto the tops of these mountains. The bare, black rocks were now dusted with white. Beyond these mountains lay the Pacific Ocean and the desert coastline upon which were built the major cities of the country.

The mayor turned toward the east and watched the light of the setting sun play on the sides of the White Mountains. He stopped for a moment to let his mind drink in the beautiful sight. Everywhere the sunlight twinkled brightly on patches of snow and ice. Towering above all, less than ten miles away, was the white-capped peak of *Nevado Huascarán,* the tallest mountain in all of Peru.

The valley is a deep, narrow gash running between these two ranges of mountains. *Nevado Huascarán* towers more than 22,200 feet above sea level, while the floor of the valley is only 9,000 feet in elevation—nearly 2½ miles below the icy peak!

People have lived in this valley for a long, long time. Farther down the valley stand the ruins of the ancient Temple of Chavín, built nearly 3,500 years ago. Its walls were covered with carvings by some forgotten race of people. On the walls are pictures that tell a story of a time when the earth and the sky were at war. According to this legend, the beautiful *Corridor of Greenery* was a slashing wound suffered by the earth during this battle.

But scientists tell a different story of the creation of this valley. They say that it was carved by the swiftly flowing water of the Santa River. Arising in a lake that is fed by melting snow, the Santa tumbles in a mass of white foam over the rocks as it flows northward through the valley before it turns westward to the ocean.

The valley is only 750 miles south of the equator and summer occurs in January. Groves of royal palms grow everywhere along the bottom of the valley, and the unpaved, ancient roads are lined with eucalyptus trees. On either side of the river, the valley is narrow and slopes quickly upward toward the foothills of the mountains. From the air, the land looks like a patchwork quilt, with tiny fields of irregular shapes separated by dark bands of trees and shrubs or white bands of stone fences.

January 10, 1962, had been just like so many other days in this lovely place. People had worked, played, laughed and enjoyed the warm sunshine. But ten miles to the east, near the top of *Nevado Huascarán,* things were not going as they usually did. At the crest of the mountain lay the sprawling mass of a glacier. Such huge hunks of ice dot the tops of so many of the White Mountains that the scientists who study them have given them numbers

rather than names. This was Glacier 511. A few of
the Inca Indian families were even now making
plans to visit Glacier 511 in a few weeks to chop ice
from its frozen surface. The women would then
carry the heavy loads to the towns in the valley to be
sold for use in the ice boxes of the rich.

But for several days the hot summer sun had
shone down on the ice of Glacier 511. The surface
of the ice was broken and cracked by the heavy load
of newly fallen snow. Into thousands of these cracks
the water of the melted snow flowed in little
streams. The melted water trickled slowly into the
openings and, when it reached the solid rock below,
seeped downhill under the glacier. Suddenly, at
6:13 in the evening, a block of ice, weighing
perhaps as much as three million tons, broke loose
from the mountain peak and slipped down the face
of the glacier.

The noise and the shock of the falling ice was
heard and felt in the valley below. A growing cloud
of ice and dust was seen clearly. But it was too late.
Those people in the path of the avalanche were
doomed to die.

Following the trough of a river canyon, the mass
of snow and ice roared down toward *Callejón de
Huailas* and its helpless people. As it went, it
gathered up tons of rocks from the walls of the
canyon, adding their weight to its own. Within two
minutes after it started, the avalanche crushed the
village of Pacucco. Moments later it smashed its way
through another village, Yanamachico. The Indian
men who had gathered in the villages to drink beer

and talk about their sheep and crops were killed
before they realized what was happening. The
women of the villages died as they cooked their
evening meals. Children were crushed while they
played. Of the more than 800 people who were in
these villages, only eight survived. The front edge
of the avalanche must have been moving at nearly
100 miles an hour at this time.

A minute later, the churning, tumbling torrent
reached the valley floor and began to spread out
like a huge fan. As it did, its speed slowed a little.
But it still bore down on Ranrahirca at more than a
mile per minute.

The mayor of Ranrahirca had stopped by the
power station to chat with the electrician and to
watch the lights of his town come on as the switches
were thrown. Above them, the millions of tons of
ice and rock bounced from one side of the narrow
canyon to the other. A few people in Ranrahirca
realized what the sounds meant. Their first thought
was to warn their friends and families. Each of
them rushed through the streets shouting warnings
as the 40-foot-high wall of ice and rock ground
toward them. Hundreds of people fled to what
they hoped would be the safest place—the church.

The electrician left the power house and pushed
his way through the crowd that rushed toward the
church. Because of the number of people packed
into the narrow street, he soon realized that he
would never reach his home in time to warn his
family. Then he saw two small girls who were
neighbors of his. They were struggling to stay on

their feet in the mass of pushing, frightened people. The electrician grabbed a child in each hand and tried to pull them to the safety of a quiet street. But as he did so the avalanche rushed down on the town. The tumbling edge of the ice snatched the girls from the man's grasp. Dust filled the air, blinding and choking him. When he could again see, the children were gone. Their bodies were never found.

The church was also gone. As the people inside prayed, the sheet of ice, that was more than twice as tall as the church's steeple, crushed them all.

The mayor of Ranrahirca was only a few blocks from his home, where he lived with his sister. As the roar of the avalanche filled the air, he turned and ran, shouting a warning to everyone he met. He reached the street in front of his house and shouted to his sister. But the noise of the ice smashing through the town drowned out his words. The thundering ice smashed past, ripping a corner from the house but not touching either the mayor or his sister.

The noise of the avalanche drowned out the warnings shouted by other people. One man heard the sound of the crashing ice while he was on his way to a party being given for his aunt. He rushed to the house where the party was being held and shouted a warning to the people inside. But no one heard him. As the 40-foot-high wall of ice loomed into sight, the man gave one last shout and then ran for safety. Twenty-seven members of his family were crushed in the ruins of the house.

The children at the birthday party were also buried before they could run from the house. The tired business man and his family were trapped by the collapsing roof of their house and died under the tons of ice. People walking down the narrow streets were caught by the ice that moved four times faster than a person could run. Of the 2,500 people who lived in Ranrahirca, only 89 survived.

The front edge of the mass of ice and rock passed on over the town and crossed the Santa River. There, 9½ miles from its starting place and 2½ miles lower than the peak of *Nevado Huascarán,* it ground to a stop. The river was completely dammed and the water backed up until it was more than 15 feet deep. Then the ice melted enough to release the flood, and two bridges downstream were washed away.

The smashed valley below the avalanche was now completely cut off from the outside world, or so the survivors thought. As they stood staring helplessly at the 40-foot-thick sheet of debris that covered their homes, a government helicopter fluttered overhead. Within hours, help was on the way. Soldiers landed in helicopters while medical supplies, doctors, nurses and even a portable hospital were brought into the small airport in large airplanes.

But the medical teams found little to do at first. They discovered that there were few injured people who needed medical aid. The avalanche had killed everyone in its path and had not touched those lucky enough to be standing to one side or the other.

But with so many dead people buried under the rapidly melting ice, along with perhaps as many as 10,000 animals, the threat of disease was very real for the people of the valley. The doctors quickly set up centers and began giving injections to the survivors to prevent the spread of typhoid. They also ordered the area sprayed with DDT in the hope that they could avoid an outbreak of typhus.

Over the next few weeks the ice melted, leaving tons of rock, dirt and smashed houses lying on the remains of Ranrahirca. Bodies were dug from the debris, identified and reburied. Refugees were fed, clothed and housed. Life slowly returned to normal. But the scar left by the avalanche will remain for many, many years, both on the face of the land and in the minds of the people who survived it.

CHAPTER NINE

THE GREAT GLAZE STORM OF 1951

"HOW MUCH FARTHER?"

"Only about fifty miles, dear," the driver of the car responded.

"Can you see OK?" his wife asked.

"Well, it seems to be raining harder. That's why I slowed down to forty."

"For goodness sakes, be careful. I'm sure we would all rather get home a little late than have an accident."

"Yes, dear," replied Mr. Turner. He was a bit disgusted since he was already driving as safely as he knew how.

It had been a tiring trip for the Turners. They had left Nashville, Tennessee, on Thursday in the late afternoon and had driven straight through to Columbus, Ohio. That way the girls didn't have to miss too much school.

On Saturday afternoon Mrs. Turner's sister had been married. And on Sunday morning they were headed back home. Before the weather turned bad, they had hoped to make it by nine o'clock. Now it looked like it would be closer to midnight.

"What time is it?" Joanne Turner asked next.

"Almost exactly ten o'clock."

"Would it bother you if I turned on the radio? Maybe they will say something about the weather. As hard as it is raining, there could even be floods."

"No, go ahead," Ned Turner answered.

The radio cracked with static and then the announcer's voice filled the car, ". . . and here are tomorrow morning's headlines:

In Korea, the U.S. Eighth Army fought to within eight miles of Seoul.

In Washington, President Truman met today with French Premier Rene Pleven.

In sports, Kentucky's once-defeated Wildcats retained their ranking today as the nation's number one basketball team."

"Mommy, are we home yet?"

"Not yet, Kerry," her mother answered as she turned to look in the crowded back seat. Still asleep were older sisters Kim and Kathy.

"How much longer?" Kerry asked for the thousandth time.

"Not much longer. An hour or so if the rain doesn't turn to ice," her father volunteered.

"What did they say about the weather?" asked Mrs. Turner, who had stopped listening to the news when Kerry woke up.

"I didn't catch it all, but they say colder air is moving in. We are likely to get ice or snow," Ned reported.

The Turner family pulled into the driveway of their Nashville home at almost exactly the stroke of midnight. It had rained hard. The closer they got to home, the slower they had to drive. It seemed as if the last ten miles had taken forever.

The weatherman was right about the cold weather. The Turners were no sooner in bed than the rain turned to ice.

With the first sound of the frozen rain against the bedroom window, Mrs. Turner sighed, "Thank goodness we're home." Then she and her husband both fell asleep, exhausted from the long trip.

The Turners all slept late on Monday morning, January 29, 1951. It was just as well since Nashville and all of central Tennessee was covered with a layer of ice. During the night the temperature had dropped from near 60°F to below freezing. But the worst was yet to come.

A great mass of polar air from Canada had swept into the central part of the United States during the last week of January. It was met in the southeastern part of the United States by a tropical air mass from the Gulf of Mexico. Warm, moist air meeting colder air often spells trouble. In this case, it was rain, and later glaze, sleet and snow. A belt more than 100 miles wide, extending from Louisiana northeasterly to West Virginia, was caught in the South's worst winter storm. And Nashville, the nation's country music capital, was right in the center of the action.

On Monday night the inch-thick glaze of ice was covered with a fresh layer of sleet and snow. Travel was brought to a standstill. Telephone and power lines were down. Thousands spent the night in the darkness and cold. Without electricity, there were no lights and many home furnaces would not operate. The mercury sank to an all-time low of 13°F below zero.

Things didn't improve much on Tuesday or Wednesday. Hundreds worked to repair the downed telephone and power lines, remove fallen trees and clear the streets. On Wednesday evening the skies were cold, gray and sullen. Another low pressure front was moving to the south, and that night there was more ice and snow. The next morning there were eight inches of frozen precipitation on the ground.

Only a few cars with chains were able to move. Everywhere there were stalled cars and trucks. Many tourists, bound for Florida vacations, were stranded, some for four or five days. Trains were stopped completely or ran late. Airplanes flew overhead, but none landed at the Nashville airport

Many tons of ice bend and break telephone wires and trees throughout New England after an ice storm.

for six days. Nashville was trapped. It would be nearly ten days before things got back to anywhere near normal.

The main reason for the difficult traveling conditions was the way the snow and ice covered the ground. At first the four or five inches of water-soaked sleet on the ground were not frozen into a solid sheet. Instead, this layer was covered with about four inches of dry snow. When temperatures fell, the snow and slush began to freeze from the top down. The wheels of vehicles would sink down to the wet pavement and spin.

Before many of the streets could be cleared, the white covering next to the ground froze into a solid sheet of ice about four inches thick. There was still some dry snow on the top, which made the streets and sidewalks all the more slick. After three or four days of daytime melting and nighttime freezing, this became a semitransparent layer of solid ice.

At first it was a big adventure for the three Turner girls to be snowbound, or perhaps it is more correct to say "ice-bound." With schools closed, they were able to catch up on their sleep. It was fun eating by candlelight, although they missed their favorite programs on television.

As the week wore on, the girls became bored. Sliding on the hill in back of the house lost most of its appeal. The ice-covered trees and shrubs lost their beauty and fascination. Checkers games by the fire often ended in arguments.

On Friday, February 2, the two older girls had a

real "knock down and drag out." One claimed that since the groundhog had not seen his shadow, winter was over. The other said that was just a silly superstition. She quoted her science teacher to support her side of the argument. This was countered by what Grandpa Turner had said about the groundhog while they were in Ohio.

Under normal conditions the two would have been sent off to their rooms to "cool off." In this case, their bedrooms, with temperatures barely above freezing, were a bit too cold. So their mother restored the family peace with popcorn, made in the fireplace, and a big bowl of apples.

Ned Turner went out to the car twice a day to listen to the news on the car radio. "Evidently, a lot of people are worse off than we are," he reported to the family. "At least with the fireplace we can keep warm. And that pile of wood in the garage is looking better all the time."

"I don't know about you, but I'll be glad when the electricity is back on," his wife replied. "I'm beginning to feel like a pioneer woman on the Oregon Trail. I don't know how many more meals I can make myself cook in this fireplace."

"Let's be thankful for the freezer full of food in the garage. Since it's out there, we don't have to worry as much about its thawing out even though the power is off." This was Ned's way of answering the complaint about fireplace cooking.

Despite thousands of telephones being out of service, officials in City Hall were "snowed under" with calls from victims of the storm. On Thursday one of the newspapers ran a list of helpful suggestions for meeting the emergency:

> Don't drive your car, even with chains on the tires, unless it's an absolute emergency. Then carry a shovel with you.
> Don't try to move fallen wires.
> Don't overheat your house.
> Don't try to thaw out frozen pipes.
> Don't have your basement pumped out today—you may have to have the same thing done again tomorrow.
> Don't make unnecessary telephone calls.
> Don't walk or stand under trees.
> Do keep children and elderly people indoors.
> Do inform authorities of fallen trees, broken live wires and other obstructions which might be dangerous.
> Do be patient. It will only be a couple of days until things are back to normal.
> And, above all, don't worry!

Walter Stirling must have been one of those Ned Turner referred to as not being as fortunate as some. What happened to Stirling, who worked at a service station, shouldn't have happened to a dog.

On late Sunday night, Stirling's car skidded on the ice and got stuck. He had to call a competitor for a wrecker to tow him to safety. On Monday night, the same thing happened. On Tuesday night, he again skidded on the ice. This time he left the car and walked home.

Wednesday morning he discovered that an ice-laden limb had crashed across the car. Besides that, one of the tires was flat. But that's not all. When he went home for lunch, he found the electric power off in his house. The gas furnace was not working. Also, the rain had leaked through the roof and was running down one of the walls.

When Stirling started back to work, he slipped on the ice on the porch and fell down the steps.

"I'm lucky to be alive," he told one of the men at the service station.

"Without bad luck, you wouldn't have any luck at all," his friend replied.

Much of the rest of the United States was having cold weather while Nashville was having its great glaze storm. In Lone Rock, Wisconsin, the temperature got down to −42°F on February 3rd. In Jacksonville, Florida, it was a chilly 29°F. Two inches of snow fell in St. Augustine, Florida, for the first time in forty years. There was snow as far south as Tampa and St. Petersburg. Nobody could remember that happening before.

The storm's effect on Nashville and the rest of the South was great. About 25 persons lost their lives and another 500 were injured in accidents associated with the storm. The U.S. Weather Bureau estimated the damage at approximately $100 million. This was far greater than the damage from any other storm to strike this part of the country, with the possible exception of damage done by hurricanes along the coast.

The cold wave damaged the peach crop in Illinois and Michigan. Also affected was the tobacco crop in Georgia and the Carolinas. Nor did the citrus groves in Florida go untouched. In addition, more than 250 head of cross-bred Brahman cattle, developed to withstand Florida's semitropical heat, were killed by the cold weather.

The 1951 ice storm was a record breaker because it was so severe, lasted so long and caused so much damage. However, ice storms are a rather common occurrence during the winter months.

Ice storms, of course, are not limited to just the southern part of the United States. The worst ice storms occur in an L-shaped belt extending from Texas northward to Kansas and then eastward across the Ohio Valley to the New England and Middle Atlantic states. Severe ice storms also occur in the northwestern part of the country.

November, 1921, was marked by two severe ice storms in opposite ends of the country. From the 19th to the 21st, the Pacific Northwest was locked in the grip of rain, sleet and snow. Portland, Oregon, was near the center of the area that had severe icing. The precipitation from this storm was particularly heavy. As many as 13 inches were recorded east of Portland in the foothills of the Cascade Mountains.

An ice storm in New England a week later was the worst one in the memory of the people living there. More than $5 million worth of damage was done to telephone, telegraph and power lines. There was at least that much damage or more to the forests.

Freezing rain or freezing drizzle is rain or drizzle that occurs when the surface temperature is below 32°F. The rain or drizzle falls in liquid form. However, it freezes upon impact, resulting in a coating of glaze, or smooth transparent or translucent ice. When this happens, we have an ice storm.

This coating of ice may vary from a thin glaze to deposits an inch or more in thickness, and in a few cases much thicker. Wires in northern Idaho during an ice storm in January, 1961, had ice deposits eight inches in diameter. In the case of the 1921 ice storm in New England, the ice on the wires between telegraph poles was estimated to weigh two tons. An evergreen tree 50 feet high may be coated with as much as five tons of ice during a severe storm.

Ice storms are usually accompanied by sleet. Sleet is ice pellets or frozen raindrops which bounce when they hit the ground. Sleet does not stick to wires and trees. However, severe sleet storms can cause hazardous driving conditions.

As in the case of other natural catastrophes, the best protection against ice storms is to know ahead of time if they are likely to occur. The National Weather Service uses such terms as *winter storm warning, glaze warning, travelers' advisory* and *ice storm* in warning the public that a freezing rain, sleet or freezing drizzle is forecast.

Since over 85 percent of the ice-storm deaths are related to traffic accidents, people are urged to stay off the streets and highways during and after an ice storm.

In many ways severe ice storms affect us more than our ancestors a hundred years ago. We are very dependent upon electricity, telephones and other means of modern communications. We also are always on the move, and without our automobiles, we are almost helpless. Ice storms are natural enemies of safe highways and wires strung between poles. As a result, power and communication companies spend millions of dollars each year pruning trees and replacing old wires and poles with stronger ones. More and more lines are being placed underground, where they are safe from the destructive force of an ice storm.

Despite the progress that is being made, an ice storm can still change life in a modern city to the way it was in a log cabin many years ago. If you don't believe it, ask the Turners in Nashville. The Turner girls are married now and have families of their own. However, every year on February 1, they celebrate the Great Glaze storm of 1951 by turning off the lights and eating by candlelight the dinner cooked in the fireplace.

CHAPTER TEN

AVALANCHE IN THE ALPS

THE EUROPEAN COUNTRY of Austria is shaped roughly like a pan. The handle of this pan points west, toward Switzerland. Most of the country is covered by the towering peaks of the Alps.

In the westernmost tip of the Austrian panhandle is the tiny province of Vorarlberg. The province is only twelve miles long and five miles wide. Here the mountain ridges run north and south, at right angles to the winter winds.

Vorarlberg has always been the victim of avalanches. Legend and history tell of hundreds of times when snow-slides crashed down the sides of the mountains, burying everything in their paths. But the worst disaster in Vorarlberg's history was a two-day period in early January, 1954. Within the borders of the little province, 388 avalanches were seen and reported. No one knows how many more went unobserved.

The worst of these hit the village of Blons. Only 367 people lived in the little town. The 90 houses they lived in were scattered without any pattern across ten square miles of valley floor. The people were farmers, and they felt their houses had to be near their fields.

Blons was divided into three districts. Church Village was the most important district. The church, vicarage, village offices, police station, two general stores and two inns were all located here. It took an hour to walk from the church to the farthest house in the village.

A twenty-minute walk away from the church stood a cluster of houses called Walkenbach. This was the geographic center of Blons. In Austrian villages, the church is usually built in the center of the village. But many years ago it was decided that the spot at the base of the 6,070-foot tall Montclav peak would be a safer place for the church in Blons. Walkenbach lay at the foot of Falv peak, which stands 6,111 feet above sea level.

A narrow, deep gorge, called Eschtobel, separates Walkenbach from Church Village. A narrow wooden bridge carried the single dirt road across the gorge. The gorge empties into the Lutz River, in the bottom of the main valley.

A third cluster of houses stands up the Lutz River, on the other face of the Falv Mountain. Three steel cables connect the widespread parts of Blons. These cables were used as an overhead railway to carry food, clothing and other goods rapidly from place to place.

Avalanches have always threatened the people living in Blons. Most of them come from the peak of the mountain called Falv. The earliest description of an attempt to protect the village from destruction was written five years after the discovery of America. In the winter of 1497 an avalanche slid down the mountain, killing ten people and many cattle. As a result, it was decided that the forests on the mountain slopes should not be cut. It was hoped that the trees and undergrowth would stop the sliding snow.

For many years, the main defense against avalanches in the Alps has been the planting and protecting of trees. This practice is still common all over Austria and Switzerland. But during the last 50 years, many other methods have been tried. Avalanches are intentionally set off before the snow becomes dangerously deep. Roads are often covered with roofs of reinforced concrete and wood. Open stretches of road are sometimes protected by thick stone walls.

The same techniques could be used to protect houses, but such construction is expensive. Some house owners try filling in the space behind their homes with rocks. The pile of rocks gives the snow a level place to slide out onto and over the roof of the house. Or, huge cement and rock walls can be built in the shape of arrow heads, pointing up the slope. It is hoped that the slide will be split into two sections and go around the house below the block.

But the people of Blons were poor. They had not been able to build the structures they needed to protect their village. They did not cut the trees on the slopes between their houses and the tops of the mountains, and they prayed that the avalanche would miss their homes.

Both Falv and Montcalv had been planted thickly with trees. But this had been many years ago. The trees had become old and sickly. In 1951, avalanches on Falv Mountain destroyed many of the trees. This set the stage for the disaster that was to come three years later.

The autumn of 1953 had been beautiful in Vorarlberg. During the first part of December only the highest peaks showed the white of snow. On December 19 snow fell in the high altitudes, but the meadows of Blons were brown and clear. The day after Christmas large, fluffy flakes reached the ground of the valley but quickly melted. By New Year's Eve there was only a foot or so of snow on the ground. It was very cold, but little new snow fell for several more days.

Most of the people in Blons enjoyed the beautiful weather. But some of them were worried. "It has snowed itself very badly," they said. They knew from bitter experience that the first snow of the year is often the one that determines whether or not an avalanche will form.

If the first snow falls on wet, warm ground early in the year, the damp flakes will partly melt and then freeze tightly to the soil. But the snow that fell during late December and early January fell on frozen ground. The particles of snow packed down into tiny crystals of ice. These formed a slippery surface that would be unable to hold the snow that would soon follow. It had, indeed, "snowed itself badly" on the slopes above Blons.

On the 8th of January the winter snows finally began. The snow danced through the air to the delight of the children and the owners of ski resorts. By Saturday morning, January 9, the entire province of Vorarlberg was snowed in. By that evening, the storm turned into a blizzard.

Many people left early for church the next morning. They trudged through two feet of snow, some of them for more than an hour to get to the nine o'clock Mass on time. After the service, the village people stayed for a while, talking about the snow on the sides of the mountains. Some people reported that the drifts stood deeply at the higher elevations. Others said that the paths were already covered and almost impassable in all parts of the valley. They talked in quiet voices, because they all knew that even a loud shout could start the snow moving down the sides of the mountains.

No one went to the inns as they usually did before returning home after church. Everyone seemed to feel that he had to get back home as quickly as possible. Those who lived near the church arrived home in time to hear a special Avalanche Warning System broadcast at noon. The report warned that more than 50 centimeters (20 inches) of new snow had fallen during the past 24 hours and that the wind was blowing very hard at high altitudes. The report warned that "there is now increased danger of slab avalanches in southern Vorarlberg."

A slab avalanche is one that starts when a huge section of hard packed snow and ice breaks off and begins to slide down the mountain slope. The mass of material in the avalanche moves on the ground. Since the trees planted on Falv and Montcalv usually turned or stopped slab avalanches, the people of Blons were not very worried.

But the snow high on the sides of Falv and Montcalv was not wet and hard packed. Near the tops of the mountains, the temperature was 5°F to 10°F below freezing. The snow remained light and fluffy. Its base stayed frozen and loose. The trees hung heavy with the white stuff.

No one saw it happen, but we can guess that a small handful of snow fell—perhaps from one of the cliff faces of Falv. Or maybe it fell from the branch of one of the many ancient trees that still stood near the top of the mountain. At exactly 9:36 on Monday morning a mass of powdered snow began to move down the steep slope.

The people of Blons had finished breakfast and were beginning their daily chores. The blizzard still raged and a thick fog blanketed the valley floor. A few people were out in the snow, shoveling paths or pushing their way toward a neighbor's house. A strong wind suddenly blew the fog away. A terrible roar filled the air. The people who were outside looked up in horror as a huge cloud of powdered snow rushed at them through the blizzard.

This was not the slab avalanche that the radio had warned them to watch for. This avalanche was one of the more deadly dust avalanches. The dry, powderlike snow had been whipped up into the air as it slid down the side of Falv. It soon hung like a cloud hugging the ground. The cloud rushed down the steep slope at more than 100 miles an hour. A tremendous wind howled ahead of it as the air in the valley was forced out of the way by the advancing snow. After the avalanche had passed, the air rushed in to fill the vacuum. This reverse wind carried snow and many pieces of debris back up the hill.

Very few people in Church Village heard the sound of the avalanche, perhaps because of the thick fog. Their first sign that anything had happened was a sudden, high pitched whine. The sound came from the shed that held the wheel at the end of the cable railways. As the people watched in amazement, first one and then the other of the thick cables were pulled through the wheel, their broken ends whining and whipping through the air.

"Something's happened on the other side of Eschtobel gorge," someone said. "Something has cut both of the cables!"

"Avalanche! It must have been an avalanche!"

"Listen! Someone is calling for help!"

"Ring the church bell! Call everyone together!"

As the bell rang, men began to gather at the church. Quickly a group of about a dozen were plowing through the snow toward the bridge over the Eschtobel gorge. At the bridge they met a tiny group of people struggling through the blizzard toward them. Walkenbach was gone, the survivors said. Everything from the peak of Falv to the river had been covered by snow or destroyed by the wind.

The rescue team crossed the bridge and found themselves standing on the edge of nothingness. For as far as they could see in the fog, the earth was covered with a thick, rough blanket of snow. Here and there a broken tree, a ruined house top or a haystack stuck up through the snow. The only living things to be seen were a few cows that had somehow managed to survive the avalanche.

The men did not know where to start. They had only five shovels, one stretcher, and a few blankets. They knew that many people buried under the snow were probably still alive, but they did not know how to find them. They had no maps to show where the houses had been. The old familiar landmarks were either gone or were covered by the snow.

The blizzard blew all day as the men dug through the shattered village. They soon discovered that few people had been carried downhill by the avalanche. Most of the victims were found directly under the debris of their houses. The wind had apparently collapsed the buildings, and the snow had done little harm other than to cover them over.

Many people who were buried in the snow but still conscious, saved themselves by calling for help at just the right time. When a voice was heard from under the snow, the rescue team could dig that person out quickly. The living were taken across the gorge to the nearest houses, which had been turned into hospitals. However, the injured people found that their problems were just beginning. The little village had only one doctor, and he had been visiting a patient in a nearby village when the storm began. No one knew where he was and there was no way to send for him. The women of the village had to take care of the injured people without the help of a trained doctor. They had no medicines at all and only a few bandages and splints. A person who was badly hurt could do little except lie in front of the fire and wait.

As darkness fell, the search for the injured had to be abandoned until morning. The rescuers knew that dozens of people still lay under that field of snow somewhere, but they had no lights to help them see. So they all gathered at the church where the priest held a brief service for the dead and dying.

They then returned to the houses that clustered around the church. The injured were made as comfortable as possible. Everyone who could talk told the others about the "miracles" that had saved them from the white death.

The mountain called Montcalv rose 3,000 feet above Church Village. Its peak was covered with old spruce and fir trees. By seven o'clock in the evening of January 11, 18 feet of very dry snow had piled up between the tall trees.

The avalanche started in the forest that was to have protected the village below. Somehow, suddenly, the entire face of the mountain began to move. Along a front, hundreds of yards wide, the snow slipped quietly through the trees. As it moved it picked up more and more speed.

By the time the mass of snow reached the younger, healthier trees on the lower slopes, it was moving too fast to be stopped. The cloud of dry snow smashed through the forest. Sometimes an entire tree was torn up by its roots. Other times only the crown of the tree was pulled off.

Because this was a dust avalanche it did not exactly follow the surface of the ground. Ridges, hills, gullies that might have changed the direction of a slab avalanche had no effect on this avalanche. Everything in its path was smashed—first by the wind, then by the snow, and again by the wind rushing in behind the avalanche. Scientists later figured out that the front of the avalanche must have been traveling at a speed of more than 200 miles an hour when it smashed through the Church Village part of Blons and into the Lutz River.

The only part of the village that escaped the second avalanche was the cluster of houses behind the church. The hill on which the church stood split the avalanche into two arms. These flowed around the church, the vicarage and a dozen other buildings. Except for these and a few buildings on the slopes, Blons was completely destroyed.

Those people who could do so struggled out into the blizzard again. They all gathered at the church. As nearly as they could tell, the avalanche from Montcalv had buried another 40 people. Sixteen of these were people who had been caught by the Falv avalanche earlier in the morning.

The snow continued to fall from the sky all through the night. It was difficult to work in the darkness and the cold, but the few people who were able to walk could not ignore the screams and pleas for help from the buried and injured people. They worked through the night. They prayed for help, but no one came. The valley was cut off from the rest of the world. No one knew of the disaster that had smashed Blons.

At dawn on Tuesday morning, three young men set out to get help. The blizzard was so fierce and the fog so thick they couldn't see more than five yards in front of them. The road had disappeared under a thick layer of snow. Above them, on the slopes, hung tons of snow waiting to slide into the valley and crush them. In spite of the fact that they had walked this road hundreds of times before, they missed the first town entirely. Hours later they stumbled into the police station in the village of Thuringen. It was almost exactly 24 hours after the first avalanche that the outside world learned of the disaster in Blons.

A group of volunteers immediately began gathering rescue equipment. It was nearly dark before the first of the rescue teams brought lights, snow probes, medicines and bandages into the ruins. In

response to calls for help broadcast over the province radio station, hundreds of well-trained avalanche rescue teams began to move toward Blons.

Wednesday morning dawned clear and bright.
Military and police units pushed their way through
the clogged road and into the village. With trained
efficiency, radio communication with the outside
world was quickly made, a spot for landing helicopters was cleared and a field hospital was set up.

Even though it had been more than 48 hours
since the first avalanche hit the village, the rescuers
began to dig in the rubble searching for survivors.
They used dogs to sniff out the scent of buried
human beings. They used 18-foot-long metal
probes that were gently pushed through the snow,
feeling for the touch of soft flesh. When these
failed, the rescuers began to dig long trenches
through the snow. Of the 40 people buried in the
Monday night avalanche, 19 were finally taken
from the snow alive.

The last person to be dug out alive was a young
shepherd. He had been in the back of his stable
feeding his animals when the snow rushed down
the side of Montcalv. The wall of the stable had
protected him from the worst of the avalanche, but
he had been buried in a mixture of straw and snow.
He lay there for nearly 62 hours before rescuers
heard his feeble calls for help.

For its size, few communities in the world have
suffered like the village of Blons. A total of 115
people—almost one in three—had been buried in
one or the other of the avalanches. Of these, 64

were dug from the snow alive, although eight of these later died. As many as 47 were dead when they were removed from the debris. And two have never been found.

More than a third of the houses in the village were destroyed along with fifty stables. Also 270 head of cattle died.

Many of the younger survivors decided to leave the valley forever. But 15 new houses and 24 stables soon were started in the village. Many of these were built at the same sites where the destroyed buildings had stood. "A man must be near his land," the older people say. "Perhaps next time, the avalanche will miss us."

CHAPTER ELEVEN

THE
SCIENCE
OF
WINTER
STORMS

DURING THE WARM DAYS of summer, the sun stands high in the sky. As the earth moves on its year-long trip around the sun, its tilted axis causes the angle of the sunlight to change. On the first day of autumn (about September 23) the sun hangs over the equator. For the next six months, until the first day of spring, the rays of the sun strike the United States at a sharper angle. Because of this, the sun's heat is spread out more in the winter than it is in the summer.

Perhaps you have seen this idea demonstrated. A flashlight is shone down on a table top. It forms a small, round circle of light. If the flashlight is tipped, the light hits the table top at an angle. It is now spread out over a large area. The spread-out light is not as bright as it was before. The amount of light and heat hitting any one spot is much less.

Of course, this is only part of the explanation of why we have seasons. The way the earth revolves around the sun with its axis tilted results in the Northern Hemisphere having more hours of sunlight in summer than in winter. The longest day occurs on about June 21. From then until around December 21 the days get shorter. The parts of the earth having the fewest hours of sunlight receive less heat energy. As a result, those areas slowly cool off, and winter weather sets in.

For the northern parts of the United States and near the tops of the high mountains, winter storms are a threat from September through May. During the four coldest months of the year (December, January, February, and March) as many as 35 violent winter storms may smash at various parts of the country.

Winter storms may take many forms. They often bring very cold temperatures. Heavy snows may fall. Freezing rain or drizzle can coat everything with a glaze of ice. Sleet and hail often pound the earth. Or all of these may hit us during the same winter storm.

While no two winter storms are exactly the same, they all can cause misery and death. Thousands of square miles of land can be cut off from the rest of the world. Millions of living things, including human beings, will suffer and often die before help can reach them. Even in the warmer areas, where it rarely snows, a sudden drop in temperature can cause the loss of millions of dollars worth of crops.

Even the fate of a country can depend upon winter weather. Several times during the American Revolution, winter storms helped the American Army, and in other battles the weather helped the British. Russia was saved twice by winter storms. Napoleon had to retreat from Moscow after the Russians burned the city in 1812 because he knew he could not survive the winter in the open. In 1943–44, winter storms again came to the aid of the Russians when the Germans tried to capture Moscow and Leningrad.

Most of the storms in the United States move from west to east. Some winter storms start off the coast of Asia and travel across the Pacific. These storms can strike the coast anywhere between Alaska and Southern California.

Most of the moisture in these western storms is dropped along the coast or on the western slopes of the mountains. This is because the damp air is forced upward by the slope. As it rises, the air cools. The cold air cannot hold as much moisture as it did when it was warmer. The extra water falls as rain or snow. These storms can strike without warning. In 1846, a storm, that was probably caused in this way, trapped the Donner wagon train. In 1952 a modern train was caught in the same area by a similar storm. It took three days to rescue the passengers.

A few storms manage to slip over the mountains and reform. Colorado and the Texas and Oklahoma panhandle areas are the breeding grounds for some of our worst storms, such as the Blizzard of '49. These mid-western storms generally move toward the Great Lakes, across the northeast, and then off the coast of New England.

The Great Lakes themselves are often the center for the development of winter storms. Storms that form here also usually follow a track to the east.

The center of most of the big storms that move through the eastern United States take a turn to the northeast over the eastern part of the country. When the center of the storm, moving eastward, is off the coast, we usually have what is known as a "nor'easter" to those living along the coast. This storm has strong winds sometimes reaching near-hurricane force from Cape Hatteras, North Carolina, northwards. Such winter storms may leave heavy snows over much of the inland sections, westward toward the mountains. A nor'easter may hit a coastal city without much warning. The famous Hurricane Hunter airplanes of the United States Navy and the Airforce are now flying into these storms in an effort to track them more accurately.

Every winter has its bad storms. The people caught in them do not really care how widespread the storm damage is. They are concerned for themselves, their loved ones and their property. But, for the country as a whole, some winters are much worse than others. On the average, 100 people in the United States will die each winter because of storms. But the Great Blizzard of 1888 killed more than 400 people. The storms during the winter of 1966 caused the deaths of 354 people. And a total of 345 people died in the winter storms of 1958.

A winter storm does not need to dump tons of snow or ice on us to be dangerous. An unexpected drop in temperature can be as deadly as a snowstorm. Even along the warm coast of the Gulf of Mexico, temperatures of 2°F below zero have been recorded. Temperatures in New England have been as low as 30°F, 40°F and even 50°F below zero! In the western mountains, records of 50°F and 60°F below zero are common. Alaska, of course, holds the record for the lowest temperature ever recorded in the United States with a terrible −76°F.

The Weather Service tries to warn us of a sudden drop in temperature. When you hear the words "cold wave warning" in a weather forecast, you know that you must get ready for temperatures lower than normal.

But you also know that the temperature readings do not tell the whole story of a cold wave. If the wind is blowing against your unprotected face or hands, you have discovered that it feels a lot colder than it really is. Going out into temperatures that are near zero can be dangerous if a brisk wind is blowing.

Scientists who work in the far north country have a chart to help them figure the effects of the wind. They call it *the wind-chill factor chart.*

To read this chart, simply find the temperature on the top line. Then read down the chart until you

WIND CHILL EQUIVALENT TEMPERATURES (°F)

WINDSPEED (MILES PER HOUR)	CALM	35	30	25	20	15	10	5	0	-5	-10	-15	-20	-25	-30	-35	-40	-45
5		33	27	21	16	12	7	1	-6	-11	-15	-20	-26	-31	-35	-41	-47	-54
10		21	16	9	2	-2	-9	-15	-22	-27	-31	-38	-45	-52	-58	-64	-70	-77
15		16	11	1	-6	-11	-18	-25	-33	-40	-45	-51	-60	-65	-70	-78	-85	-90
20		12	3	-4	-9	-17	-24	-32	-40	-46	-52	-60	-68	-76	-81	-88	-96	-103
25		7	0	-7	-15	-22	-29	-37	-45	-52	-58	-67	-75	-83	-89	-96	-104	-112
30		5	-2	-11	-18	-26	-33	-41	-49	-56	-63	-70	-78	-87	-94	-101	-109	-117
35		3	-4	-13	-20	-27	-35	-43	-52	-60	-67	-72	-83	-90	-98	-105	-113	-123
40		1	-4	-15	-22	-29	-36	-45	-54	-62	-69	-76	-87	-94	-101	-107	-116	-128
45		1	-6	-17	-24	-31	-38	-46	-54	-63	-70	-78	-87	-94	-101	-108	-118	-128
50		0	-7	-17	-24	-31	-38	-47	-56	-63	-70	-79	-88	-96	-103	-110	-120	-128

are opposite the speed of the wind. For example, suppose your thermometer reads +5°F and the wind is blowing at 30 miles per hour. You find the column headed +5 on the top line. You follow that column down to 30 miles per hour wind speed and you read −41°F. The chill-factor of the wind turns a temperature of +5°F into a killing 41°F below zero! A person who is dressed for a +5°F temperature would quickly freeze to death in this storm.

Low temperatures, plus wind, plus snow over a period of time make a blizzard. This is the most dangerous of all winter storms. Many of the storms you read about in this book were blizzards. The Weather Service sends out two kinds of warnings for these types of storms. A simple *blizzard warning* means that you can expect winds of at least 35 miles per hour, temperatures of 20°F or lower, and a lot of snow for several hours. If you hear a *severe bliz-*

zard warning, you should get ready for temperatures of 10°F or lower, winds of 45 miles per hour or more and blinding snow.

Snow is frozen water vapor. It forms high in the air. Here the air is holding all of the moisture it can. The extra water vapor freezes into the large, white, six-sided snowflakes. If the snow falls through a layer of warm air, the flakes may join together and form huge, wet snowflakes.

If you have never been in a blizzard, try to imagine what it must be like. If the temperature is 20°F and the wind is blowing at 35 miles per hour, the chill-factor makes the temperature feel the same as −20°F on a calm day. In a severe blizzard, the air would feel the same as 38°F below zero in still air. Add to these terribly cold temperatures a heavy snow, blowing so thickly that you cannot see more than a few feet in front of you, and you have a blizzard.

Ice storms are another very dangerous kind of winter storm. If the temperature near the ground is below freezing (32°F) and the temperature of the air some distance above the ground is above freezing, you may be in for an ice storm if rain is expected.

Ice storms are caused by *warm* air. Imagine that it has been very cold and clear for several days. You are in the middle of a mass of cold air. The weather report says that a "warm front" is on the way.

"Good," you think. "Now it will warm up for a while."

But watch out! Before that warm air reaches you, you may have trouble.

Because the warm air is lighter, it rises up and over the cold air. The warm moist layer of air that is on top becomes cooler as it rises. Since the air cannot hold as much water as it could when it was warmer, clouds form. Rain begins to fall. But the temperature near the ground is still very cold. The water freezes as it hits the ground and a glaze begins to form.

Tons of ice may cover every large tree. Ice forms on electric power lines. Sometimes the weight is too great. The tree limbs break. The wires fall. Thousands of people lose their electric power and telephone service.

However the biggest danger of ice storms is on the ground. The glaze of ice will cover the sidewalks and the roads causing people to slip and fall on the ice, sometimes receiving serious injuries. Automobile accidents are common. Most of the people who die during ice storms are killed in traffic accidents.

Sleet usually accompanies ice storms. In an ice storm, the water falls as a liquid and freezes before it hits the ground. Sleet is frozen rain. Sleet hits the ground in solid form. You can tell sleet from rain because the sleet will bounce when it hits something. Sleet will not stick to wires or tree limbs. But if enough sleet collects on the ground, driving and walking can become dangerous.

In the high, steep mountains of the world, another winter danger is the avalanche. The word "avalanche" can be used to mean any type of material sliding or falling down the side of a mountain. In this book, you have read about avalanches of snow and ice.

Scientists can identify dozens of different types of snow or ice slides. There are, however, only two main types. One of these is the *slab avalanche.* A mass of snow or ice breaks loose from the mountain peak and slides down the slope. In this type of avalanche, the sliding material is tightly packed together. It changes its course when it strikes something, or it climbs up and over the object blocking it.

The second type of avalanche is often called a *dust avalanche.* In these slides, the snow is not tightly packed. Instead, it is loose and fluffy. This condition usually happens in the dead of winter, when the temperature of the air is very cold. The droplets of snow trap air as they move down the slope until the avalanche is formed by a huge cloud. The air in the valley is forced out of the way of the rushing cloud, which may be traveling at more than 200 miles an hour. People, houses, even trees may be knocked over by the wind hundreds of feet in front of the avalanche.

The main cause of avalanches is too much snow falling on a steep slope. But the scientists who study avalanches have found that the problem is much more complicated than this. How heavy the snow is, how well the snow crystals will hold together and how slippery the base of the snow bank is are all important questions. The strength and direction of the wind, the temperature and the amount of trees and undergrowth also control whether or not an avalanche will start.

Once the snow is balanced on the side of a high mountain, almost anything can start it down the slope. Even the sound of thunder, a gun shot or a passing airplane can be dangerous. It is often possible to trigger an avalanche before the snow builds up to a dangerous level. Often loud noises are used to do this. Skiers may cross a slope at just the right point and cause a small, not-too-dangerous slide to begin. Sometimes it may be necessary to fire a round of artillery into a mass of snow to get it to move.

Armies fighting in the Alps during World War I learned how to use avalanches as weapons. More than 10,000 Austrian and Italian troops were killed by avalanches on one day in 1916. During this war, at least 60,000 people were killed by slides, many of them started by enemy troops.

Engineers are experimenting with various types of anti-avalanche structures. Metal and stone embankments have been built across the paths of some of the largest avalanches in the hope that these will change the course of the slide or even stop it. Other structures have been built near the tops of some mountains in an attempt to stop the build-up of the snow there.

Several hundred people are killed by avalanches each year in spite of modern warning systems and better anti-avalanche buildings. Quite often a person buried by an avalanche will die very quickly. Slab avalanches usually crush their victims. People caught in dust avalanches generally smother to death in the powdery snow. But many times, a person buried in an avalanche is lucky and stays alive until he can be rescued.

The fastest way to find a person buried in the snow is with an avalanche dog. These animals, usually Alsatians or Belgian Shepherds, have been trained to use their sense of smell to find avalanche victims.

Winter storms of all kinds are dangerous. Traffic accidents and heart attacks are about equally deadly during winter storms. A few people freeze to death in some storms. Others are killed in fires caused by overloaded heating systems. Still more die in stalled cars, killed by the carbon monoxide gas produced by motors kept running to keep the people warm. Falls on the ice kill some people each year. Electric wires downed by ice are also dangerous. And, occasionally, a building will collapse because of the weight of snow and ice, trapping and killing the people inside.

Usually it is difficult to get out of the way of a bad winter storm. In most cases we must simply wait until the weather clears and the temperature goes back to normal. There are many things a person can do to make "riding out the storm" easier and safer. The most important is to listen to the warnings and advice of the Weather Service. It is difficult to predict exactly how bad a winter storm will be, but many people die each year because they did not listen to the directions of the weather scientists.

In order to keep up with the latest weather information, you will need a portable radio or television set. Remember that bad storms quite often destroy electric power lines. Your home may be without electricity for several hours, or even for days.

Of course, you will need a supply of food that will last through the storm. But without electricity, it might be impossible to cook your food. So, for such an emergency, the food should be the kind that can be eaten without cooking. Or you could have an emergency source of heat for cooking.

Staying warm during a winter storm is a serious problem. Homes that are heated electrically can lose their source of heat easily. Most central furnaces that burn fuel oil or gas depend upon electricity to operate the fan that blows the hot air through the house. Many people forget that these furnaces will not work if there is a power failure. However, if you dress warmly and stay indoors, you have a good chance of surviving, even if your house is cold.

Staying indoors during a winter storm is very important. Many people get lost only a few feet from their homes during blizzards. Shoveling snow, pushing cars, even just walking through heavy snow is terribly hard work. Becoming overly tired can bring on heart attacks in anyone who is not in perfect physical condition.

Many of the same rules apply to people who are traveling by automobile during the winter months.

The most important rule is to listen to the weather warnings constantly. In many parts of the country, a winter storm can strike quickly.

People traveling by car during the winter should always carry certain emergency supplies. The chance of being trapped for several days is always there. Before starting out on such a trip, the car should be as well prepared as possible. It should have a full tank of gas. It should be in good mechanical condition. Its heater should work well. It should have snow tires, and a set of chains should be carried. Also in the trunk should be a supply of food, blankets or sleeping bags, a shovel, a sack of sand, a camping heater, tow chains and a fire extinguisher.

If you are trapped in a car, your best bet is to stay where you are. Remember the chill-factor. If you stay out of the wind, you will be warmer and less likely to freeze to death than you would be out in the open. It is easy to get lost in blowing snow. And it is easier for rescue parties to find a car on the side of the highway than it is to find a person alone in the woods or fields.

By staying in your car you will also avoid over-exertion. Do not spend too much effort trying to dig your car from the snow bank. The danger of heart attack is always a threat to the person who is working in the cold.

In spite of the cold, a window must be left open in a trapped car. Wet snow can make a car air tight causing the people inside to suffocate. If you must run the motor of the car to keep the heater going, you face the danger of dying of carbon monoxide poisoning. Therefore, ventilation is always necessary.

Keep awake as much as possible, and exercise regularly. Clapping hands and moving your legs are necessary when you are in a cramped car for a long time. Someone should always be awake, to watch for signs of carbon monoxide poisoning in the other people and to be ready to signal to rescue crews. As soon as the storm lets up a little, people will be searching the highways. But a car buried in a snowbank along the side of the road may be overlooked.

Winter storms can cause great hardships. But what kind of a world would it be if there were no winter storms? Without the moisture of the heavy winter snows, the *breadbasket of America,* where much of our wheat, corn and other crops are grown, would become a barren desert. The number of insects is controlled by winter temperatures, and a mild winter is often followed by a larger number of insects in the summer. No snow would mean no skiing—no recreation for thousands and no income for those in the tourist business. How many other ways can you think of that your life would be different without winter storms?

ACKNOWLEDGEMENTS

Photographs in this book are reproduced courtesy of the following:

pages 2–3: California Historical Society
pages 36–37: New York Historical Society
pages 62–63, 99: Wide World Photos
pages 86–87: Library of Congress
pages 100, 120–121: National Oceanic and Atmospheric
    Administration

INDEX

About the Authors

Walter R. Brown has his Ph.D. in science education from The Ohio State University and enjoys teaching junior high school students. He has co-authored a science textbook series that is used in many junior high school science courses throughout the country. Dr. Brown lives in Virginia Beach, Virginia, with his wife and three of their five children.

Norman D. Anderson also received his Ph.D. from The Ohio State University and is presently Professor of Science Education at the North Carolina State University. He has several science books to his credit, including some he co-authored with Walter Brown. Dr. Anderson lives in Raleigh, North Carolina, with his wife and six children.